ENVIRONMENTAL SCIENCE, ENGINEERING AND TECHNOLOGY

CARS, CLIMATE AND THE EPA: ISSUES IN REGULATING GREENHOUSE GAS EMISSIONS

Environmental Science, Engineering and Technology

Additional books in this series can be found on Nova's website under the Series tab.

Additional E-books in this series can be found on Nova's website under the E-books tab.

ENVIRONMENTAL SCIENCE, ENGINEERING AND TECHNOLOGY

CARS, CLIMATE AND THE EPA: ISSUES IN REGULATING GREENHOUSE GAS EMISSIONS

CAROLYNNE D. BURGDOERFER
EDITOR

Nova Science Publishers, Inc.
New York

For permission to use material from this book please contact us:
Telephone 631-231-7269; Fax 631-231-8175
Web Site: http://www.novapublishers.com

Library of Congress Cataloging-in-Publication Data

Cars, climate, and the EPA : issues in regulating greenhouse gas emissions / editor, Carolynne D. Burgdoerfer.
p. cm.
Includes index.
ISBN 978-1-61728-963-7 (hardcover)
1. Automobiles--Motors--Exhaust gas--Environmental aspects. 2. Greenhouse gas mitigation--Government policy--United States. 3. United States. Environmental Protection Agency. I. Burgdoerfer, Carolynne D.
TD886.5.C38 2010
363.738'7460973--dc22
2010025967

Published by Nova Science Publishers, Inc. ✢ New York

CONTENTS

PREFACE

As Congress considers legislation to reduce the greenhouse gas (GHG) emissions that contribute to climate change, attention has focused on "cap-and-trade" legislation. Such legislation would set a national cap of GHG emissions, with allowances to emit limited amounts of the gases distributed or auctioned to affected parties. Recently, there has also been discussion of taxes on greenhouse-gas-emitting substances, which proponents argue would provide greater transparency and a clearer price signal. This book examines the proponents of legislation who counter that the threat of climate change is too important for action to be delayed, and that energy-efficiency and lower GHG emissions can be the building blocks of a program to restore the economy as well as to protect the environment.

Chapter 1 - As Congress considers legislation to reduce the greenhouse gas (GHG) emissions that contribute to climate change, attention has focused on “cap-and-trade” legislation. Such legislation would set a national cap on GHG emissions, with allowances (permits) to emit limited amounts of the gases distributed or auctioned to affected parties. Recently, there has also been discussion of taxes on greenhouse-gas-emitting substances (generally referred to as a “carbon tax”), which proponents argue would provide greater transparency and a clearer price signal.

Enacting greenhouse gas controls is not simply a choice between cap-and-trade and carbon tax options, however. A third set of options, using the more traditional regulatory approaches of the Clean Air Act (CAA), is available. Unlike a cap-and-trade system or a carbon tax, regulation under the Clean Air Act does not require new Congressional action. The ability to limit GHG emissions already exists under various CAA authorities that Congress has enacted, a point underlined by the Supreme Court in an April 2007 decision (*Massachusetts v. EPA*).

Thus, controlling GHGs could follow a two-track approach, with Congress and the Administration pursuing new legal authority (for cap-and-trade, carbon tax, or whatever) at the same time that the Administration, through the Environmental Protection Agency (EPA), exercises existing authority under the Clean Air Act to begin regulation of greenhouse gas emissions. The Administration has made clear that its preference would be for Congress to address the climate issue through new legislation. Nevertheless, it is moving forward on several fronts to define how the Clean Air Act might be used.

The key to using the Clean Air Act's authority is for the EPA Administrator to find that GHG emissions are air pollutants that endanger public health or welfare. Administrator Jackson finalized such an endangerment finding, December 7, 2009. With the endangerment finding finalized, the agency can proceed to regulate emissions from various sources of GHGs.

EPA has received 10 petitions asking that it make endangerment findings and proceed to regulate emissions of greenhouse gases. Nine of the 10 petitions address mobile sources, including motor vehicles, aircraft, ships, nonroad vehicles and engines, and fuels, under Title II of the Clean Air Act. This chapter discusses EPA's authority under Title II and provides information regarding the mobile sources that might be regulated under this authority. Among these sources, motor vehicles (passenger cars and light trucks, including SUVs) are the initial targets for regulation, both because a petition addressing these sources began EPA's consideration of the endangerment issue, and because these are the most significant GHG emission sources among those covered by Title II. EPA proposed GHG emission standards for new motor vehicles September 28, 2009, and expects to finalize them in time for their application to Model Year 2012 vehicles.

Regulation of GHGs from mobile sources might also lead the agency to establish controls for stationary sources, such as electric power plants. That option, the authority for which would come from different parts of the Clean Air Act, is addressed in CRS Report R40585, *Climate Change: Potential Regulation of Stationary Greenhouse Gas Sources Under the Clean Air Act*.

Chapter 2 - On December 15, 2009, the Environmental Protection Agency (EPA) took its most important action to date related to climate change. EPA published its final determination that the combined greenhouse gas (GHG) emissions from new, light-duty motor vehicles in the United States contribute to an "endangerment" from climate change. More precisely, EPA found that such emissions, in the words of Clean Air Act (CAA) section 202(a), "cause or contribute to air pollution that may reasonably be anticipated to endanger public health or welfare." Under section 202(a), this finding requires that EPA promulgate "standards" to control such emissions—as the agency proposed to do in advance of its endangerment determination.

Some groups have objected to the endangerment determination and the emission standards to follow, arguing they will trigger a "cascade" of unacceptable regulatory consequences under *other* CAA provisions. These regulatory consequences, they say, would impose unattainable GHG-concentration goals on EPA and the states, and/or economically and administratively unreasonable burdens. This chapter examines the CAA provisions that have figured in this debate to see whether this alleged cascade of legal consequences likely would occur.

First, the report examines CAA sections that, like section 202(a), are triggered by endangerment findings. Of these, the one most likely to require EPA regulatory action after the 202(a) endangerment finding is section 111, authorizing new source performance standards—but only as to stationary source categories emitting the largest amounts of GHGs. Section 111, however, affords EPA wide discretion in setting new source performance standards. Two other sections that arguably might be triggered are 108, requiring national ambient air quality standards, and 115, which requires states to revise their implementation plans to prevent or eliminate the endangerment of public health or welfare in a foreign country. As to these sections, however, the arguable infeasibility of the regulatory goals—even if GHG emissions in the United States are significantly reduced, atmospheric concentrations would decline little—will give EPA room to argue that regulatory action is not mandatory. Other endangerment-triggered sections of the CAA can be distinguished from section 202(a) by their explicit terms, and thus would likely not be triggered by the 202(a) endangerment finding.

Second, the report looks at CAA provisions having no endangerment trigger. Of these, EPA has conceded that two require the agency to act after it promulgates the required emission standards following the 202(a) endangerment finding. One provision would require EPA to impose "best available control technology" (BACT) on GHG emissions from any major emitting facility proposed to be constructed in a Prevention of Significant Deterioration area. The other, Title V, creates an operating permit program for stationary sources of emissions, and would require stationary sources subject to BACT under the first provision to also apply for Title V permits. As to each of these requirements, EPA has proposed a "tailoring rule" setting emission thresholds far higher than in the CAA, at least for a few years. EPA justifies the departure from statutory language under the case law doctrines of "absurd results" and "administrative necessity."

A caveat: the issue analyzed in this chapter is important primarily if Congress does not enact climate change legislation that puts regulation of GHGs beyond the reach of some of the CAA provisions discussed here. In particular, the House-passed climate change bill, H.R. 2454 (the American Clean Energy and Security Act of 2009), states that three of the CAA sections treated in this chapter, and one CAA title, may not be used to address air pollutants based on their climate

Chapter 3 - Transportation fuels are required by federal law to contain a minimum amount of renewable fuel each year. This renewable fuel standard (RFS), established by the Energy Policy Act of 2005 (EPAct, P.L. 109-58) and amended by the Energy Independence and Security Act of 2007 (EISA, P.L. 110-140), requires that 12.95 billion gallons of renewable fuels be blended into gasoline and other transportation fuels in 2010. Most of this mandate will be met using corn-based ethanol. However, within the overall RFS there are secondary mandates for the use of cellulosic biofuels, biomass-based diesel fuels, and other advanced biofuels. Questions have been raised over whether there is enough feedstock supply and production capacity to meet these carveouts.

The Environmental Protection Agency (EPA) has the authority to waive the RFS requirements, in whole or in part, if certain conditions outlined in the law are present. In 2008 the governor of Texas requested a waiver of the RFS because of high grain prices, although that waiver request was denied because EPA determined that the RFS requirements alone did not "severely harm the economy of a State, a region, or the United States," a standard required by the statute. In February 2010, as part of a final rulemaking implementing the RFS as expanded by EISA, EPA waived most of the 2010 cellulosic biofuel carveout—EISA set the mandate at 100 million gallons but EPA is only requiring 6.5 million gallons, more than 90% less than scheduled by EISA. EPA cited a lack of current and expected production capacity, driven largely by a lack of investment in commercial-scale refineries.

This chapter provides a brief overview of the RFS program and discusses the process and criteria for EPA to approve a waiver petition.

Chapter 4 - Alternative fuels and advanced technology vehicles are seen by proponents as integral to improving urban air quality, decreasing dependence on foreign oil, and reducing emissions of greenhouse gases.

However, major barriers—especially economics—currently prevent the widespread use of these fuels and technologies. Because of these barriers, and the potential benefits, there is continued congressional interest in providing incentives and other support for their development and commercialization.

Alternative fuels and advanced technology vehicles have were addressed early in the 111th Congress, as both the House and Senate versions of the American Recovery and Reinvestment Act of 2009 (H.R. 1) contained provisions supporting their development and deployment. While some of these provisions were removed in conference, the final version contains provisions for tax incentives, federal grants and loans, and other federal support for alternative fuels and advanced vehicles.

On February 3, 2010, the Environmental Protection Agency (EPA) finalized new rules for the renewable fuel standard (RFS) that was expanded by the Energy Independence and Security Act of 2007 (EISA, P.L. 110-140). In 2010, the RFS will require the use of 12.95 billion gallons of ethanol and other biofuels in transportation fuel. Within that mandate, the RFS will require the use of 0.95 billion gallons of advanced biofuels, including 6.5 million gallons of cellulosic biofuels. EISA also requires that advanced biofuels (as well as conventional biofuels from newly built refineries) meet certain lifecycle greenhouse gas reduction requirements. EPA's methodology and conclusions on various biofuels' lifecycle emissions have been controversial.

The 111th Congress is likely to further discuss alternative fuels and advanced technology vehicles as it addresses other key topics. These include their role in any federal policy to address climate change, and their role in federal energy policy. The 111th Congress may also play an oversight role in the development of major regulations: the Environmental Protection Agency's implementation of the RFS; the Department of Transportation's implementation of new fuel economy standards enacted in 2007; and the Department of Agriculture's implementation of a new Farm Bill enacted in 2008. Further, some key tax incentives for biofuels expired at the end of 2009, and may be extended in the second session of the 111th Congress.

Chapter 5 - In May 2009, the U.S. administration announced plans to increase the Department of Transportation's (DOT) National Highway Traffic Safety Administration's (NHTSA) corporate average fuel economy (CAFE) standards and establish the Environmental Protection Agency's (EPA) greenhouse gas (GHG) emissions standards for vehicles. NHTSA redesigned CAFÉ standards for light trucks for model years 2008 through 2011, and some experts raised questions about the rigor of the computer modeling NHTSA used to develop these standards.

GAO was asked to review (1) the design of NHTSA and EPA's proposed standards; (2) how they are collaborating to set these standards; (3) improvements compared to a previous rulemaking, if any, NHTSA made to the modeling; and (4) the extent to which NHTSA analyzed the effects of past light truck standards and the accuracy of data used to set them.

GAO reviewed relevant rulemaking and modeling documents, and interviewed agency officials and other experts.

In: Cars, Climate and the EPA: Issues in... ISBN: 978-1-61728-963-7
Editors: Carolynne D. Burgdoerfer

Chapter 1

CARS AND CLIMATE: WHAT CAN EPA DO TO CONTROL GREENHOUSE GASES FROM MOBILE SOURCES?

James E. McCarthy

SUMMARY

As Congress considers legislation to reduce the greenhouse gas (GHG) emissions that contribute to climate change, attention has focused on "cap-and-trade" legislation. Such legislation would set a national cap on GHG emissions, with allowances (permits) to emit limited amounts of the gases distributed or auctioned to affected parties. Recently, there has also been discussion of taxes on greenhouse-gas-emitting substances (generally referred to as a "carbon tax"), which proponents argue would provide greater transparency and a clearer price signal.

Enacting greenhouse gas controls is not simply a choice between cap-and-trade and carbon tax options, however. A third set of options, using the more traditional regulatory approaches of the Clean Air Act (CAA), is available. Unlike a cap-and-trade system or a carbon tax, regulation under the Clean Air Act does not require new Congressional action. The ability to limit GHG emissions already exists under various CAA authorities that Congress has enacted, a point underlined by the Supreme Court in an April 2007 decision (*Massachusetts v. EPA*).

Thus, controlling GHGs could follow a two-track approach, with Congress and the Administration pursuing new legal authority (for cap-and-trade, carbon tax, or whatever) at the same time that the Administration, through the Environmental Protection Agency (EPA), exercises existing authority under the Clean Air Act to begin regulation of greenhouse gas emissions. The Administration has made clear that its preference would be for Congress to address the climate issue through new legislation. Nevertheless, it is moving forward on several fronts to define how the Clean Air Act might be used.

The key to using the Clean Air Act's authority is for the EPA Administrator to find that GHG emissions are air pollutants that endanger public health or welfare. Administrator Jackson finalized such an endangerment finding, December 7, 2009. With the endangerment finding finalized, the agency can proceed to regulate emissions from various sources of GHGs.

EPA has received 10 petitions asking that it make endangerment findings and proceed to regulate emissions of greenhouse gases. Nine of the 10 petitions address mobile sources, including motor vehicles, aircraft, ships, nonroad vehicles and engines, and fuels, under Title II of the Clean Air Act. This chapter discusses EPA's authority under Title II and provides information regarding the mobile sources that might be regulated under this authority. Among these sources, motor vehicles (passenger cars and light trucks, including SUVs) are the initial targets for regulation, both because a petition addressing these sources began EPA's consideration of the endangerment issue, and because these are the most significant GHG emission sources among those covered by Title II. EPA proposed GHG emission standards for new motor vehicles September 28, 2009, and expects to finalize them in time for their application to Model Year 2012 vehicles.

Regulation of GHGs from mobile sources might also lead the agency to establish controls for stationary sources, such as electric power plants. That option, the authority for which would come from different parts of the Clean Air Act, is addressed in CRS Report R40585, *Climate Change: Potential Regulation of Stationary Greenhouse Gas Sources Under the Clean Air Act*.

INTRODUCTION

In the 111th Congress, climate change legislation is among the leadership's highest priorities. With the inauguration of President Obama, there is a proponent of greenhouse gas (GHG) legislation in the White House, as well, markedly improving the prospects for enacting some sort of legislation to reduce GHG emissions. The President has said that a new energy, environment, and climate policy will be "a leading priority of my presidency, and a defining test of our time."

Not all parties are in agreement that controls on GHGs are desirable. Some argue that the science behind climate change is too uncertain; they oppose controls unless, at some future time, more is known about trends in climate variables and the impact of specific pollutants. Others, who agree in principle that the science dictates GHG reductions be made, raise a number of arguments, including issues concerning international commitments (i.e., whether China, India, and other developing countries will be subject to binding agreements) and the timing of a control regime (e.g., whether a major regulatory initiative is appropriate during the worst economic climate since the Great Depression). Still others can be expected to raise objections about the form and substance of specific legislative proposals, many of which are among the broadest regulatory measures that Congress has ever considered.

Proponents of legislation counter that the threat of climate change is too important for action to be delayed, and that energy-efficiency and lower GHG emissions can be the building blocks of a program to restore the economy as well as to protect the environment.

Option 1: Cap-and-Trade[1]

As debate has proceeded, attention has focused primarily on cap-and-trade legislation[2] that would set a national cap on emissions. Under a cap-and-trade program, the government would either distribute (on the basis of historic emissions or other factors) or auction off allowances to emit greenhouse gases; a combination of these approaches, in which some allowances are distributed and some auctioned is included in both the House-passed cap-and-trade bill (H.R. 2454) and the bill approved by the Senate Environment and Public Works Committee (S. 1733). Over time, the cap on emissions and hence the number of allowances available would be reduced, forcing sources to reduce their emissions or buy allowances from an ever-shrinking, and presumably more expensive, pool. A shrinking cap and rising prices for allowances would provide incentives for utilities and energy companies to develop less-carbon-intensive energy supplies, for manufacturers to develop more energy-efficient production processes, for the transportation sector to be made more fuel-efficient, and for greener residential, commercial, and institutional buildings.

Although cap-and-trade has drawn the most attention (and appears to be the choice of most congressional leaders and the Obama Administration), others have suggested that this approach is overly complicated. It would require the development of an entirely new administrative structure. Its costs are uncertain: the experience with emission allowance markets to date suggests that there could be substantial allowance price volatility. And the markets it creates may not be transparent: it could give rise to an array of financial instruments and trading mechanisms that would reward traders and would require strong regulatory oversight to mitigate potential market manipulation.

Option 2: A Carbon (or GHG) Tax[3]

Many cap-and-trade critics suggest that a carbon (or greenhouse gas) tax may be a more straightforward approach. By imposing a tax tied to the global warming potential of various fuels and greenhouse gas emissions, there would be clear signals to utilities, manufacturers, other businesses, government agencies, and consumers, leading them to lower-carbon choices. If the tax were imposed "upstream" (i.e., on fuel producers and importers, utilities, and/or major emitters of other GHGs), the additional cost could then be passed through to consumers of all kinds.

A carbon tax has critics of its own. The principal argument against it is that it is an instrument designed to fix the price of GHG control, whereas a cap-and-trade system establishes the quantity of GHGs emitted. If the chosen price (i.e., the tax level) insufficiently reduces emissions, Congress or a regulatory authority would need to adjust the tax level, thus negating one of the tax's supposed advantages, its cost certainty. With consensus forming among proponents of GHG control that emissions need to be ratcheted down as much as 80% over a 40-year period, the tax would need a transparent escalator provision to create additional emission reduction incentives.

From a political point of view, Congress has found it difficult to impose new taxes or to adjust those in place. Yet carbon tax proponents presume that if the tax is ineffective, it will be adjusted.[4] It is also unclear how a carbon tax would mesh with the international commitments to limit emissions that are being negotiated: if a treaty is agreed to, it is likely to establish numerical limits on emissions, not specify the cost of an effective limit.

Option 3: Regulation under Existing Authority

However these arguments play out, enacting greenhouse gas controls is not simply a choice between cap-and-trade and carbon tax options. A third set of options, using the more traditional regulatory approaches of the Clean Air Act (CAA) and other legislation, is available. These approaches could be strengthened through new legislation, but unlike a cap-and-trade system or carbon tax, regulation does not necessarily require new congressional action. The ability to limit GHG emissions already exists under various Clean Air Act authorities that Congress has enacted, a point underlined by the Supreme Court in an April 2007 decision (discussed below).

Thus, controlling GHGs could follow a two-track approach, with Congress and the Administration pursuing new legal authority (for cap-and-trade, carbon tax, or whatever) at the same time that the Administration, through the Environmental Protection Agency (EPA), exercises existing authority under the Clean Air Act to begin regulation of greenhouse gas emissions.

THE ENTRY POINT: *MASSACHUSETTS VS. EPA*

The regulatory approach has been under consideration at EPA for more than a decade. In 1998, EPA's General Counsel, Jonathan Cannon, concluded in a memorandum to the EPA Administrator that greenhouse gases were air pollutants within the Clean Air Act's definition of the term, and therefore could be regulated under the act.[5] Relying on the Cannon memorandum as well as the statute itself, on October 20, 1999, a group of 19 organizations petitioned EPA to regulate greenhouse gas emissions from new motor vehicles under Section 202 of the act.[6] Section 202 gives the EPA Administrator broad authority to set "standards applicable to the emission of any air pollutant from any class or classes of new motor vehicles" if in her judgment they contribute to air pollution which "may reasonably be anticipated to endanger public health or welfare."

EPA denied the petition August 28, 2003,[7] on the basis of a new General Counsel memorandum the same day, in which it concluded that the CAA does not grant EPA authority to regulate CO2 and other GHG emissions based on their climate change impacts.[8] The denial was challenged by Massachusetts, 11 other states, and various other petitioners in a case that ultimately reached the Supreme Court. In an April 2, 2007 decision (*Massachusetts v. EPA*), the Court found by 5-4 that EPA *does* have authority to regulate greenhouse gas emissions, since the emissions are clearly air pollutants under the Clean Air Act's definition of that term.[9] The Court's majority concluded that EPA must, therefore, decide whether emissions of these pollutants from new motor vehicles contribute to air pollution that may reasonably be anticipated to endanger public health or welfare. If it makes this finding of endangerment, the act requires the agency to establish standards for emissions of the pollutants.[10]

The ANPR

In nearly two years following the Court's decision, the Bush Administration's EPA did not respond to the original petition or make a finding regarding endangerment. Its only formal action following the Court decision was to issue a detailed information request, called an Advance Notice of Proposed Rulemaking (ANPR), on July 30, 2008.[11]

The ANPR occupied 167 pages of the *Federal Register*. Besides requesting information, it took the unusual approach of presenting statements from the Office of Management and Budget, four Cabinet Departments (Agriculture, Commerce, Transportation, and Energy), the Chairman of the Council on Environmental Quality, the Director of the President's Office of Science and Technology Policy, the Chairman of the Council of Economic Advisers, and the Chief Counsel for Advocacy at the Small Business Administration, each of whom expressed their objections to regulating greenhouse gas emissions under the Clean Air Act. The OMB statement began by noting that, "The issues raised during interagency review are so significant that we have been unable to reach interagency consensus in a timely way, and as a result, this staff draft cannot be considered Administration policy or representative of the views of the Administration."[12] It went on to state that "... the Clean Air Act is a deeply flawed and unsuitable vehicle for reducing greenhouse gas emissions."[13] The other letters concurred. The ANPR, therefore, was of limited use in reaching a conclusion on the endangerment issue, and, in any event, it presents the views of an Administration no longer in power. [14]

The Obama Administration's Approach

The current Administration made review of the endangerment issue a high priority. On April 17, 2009, EPA proposed a finding that GHGs do endanger both public health and welfare and that GHGs from new motor vehicles contribute to that endangerment. [15] Publication of the proposal in the *Federal Register*, April 24, began a 60-day public comment period. In addition, public hearings were held in mid-May in Arlington, VA, and Seattle, WA.

The proposed endangerment finding was followed in a matter of weeks by an announcement that the Administration had reached agreement with representatives of the auto industry, California, EPA (which had developed its own GHG emission standards for motor vehicles), the National Highway Traffic Safety Administration (NHTSA administers fuel economy standards for cars and trucks), and others. Under the agreement announced by the President, May 19, EPA and NHTSA will integrate corporate average fuel economy (CAFE) standards for new cars and light trucks with national greenhouse gas emission standards to be issued by EPA.

The objective of the new GHG standards will be to reach reduction levels similar to those adopted by the state of California, although the specifics of the requirement may be different.[16] A formal set of proposed standards appeared in a joint EPA-NHTSA notice in the September 28, 2009 *Federal Register*.[17] Publication of the proposal began a 60-day public comment period, with public hearings in late October in Detroit, New York, and Los Angeles. The joint standards must be finalized by March 31, 2010, in order to take effect with 2012 model year vehicles.

On December 7, 2009, another step was taken with the finalization of the endangerment finding proposed in April.[18]

EPA Administrator Jackson and others in the Administration have made clear that their preference would be for Congress to address the climate issue through new legislation. In the press release announcing the proposed endangerment finding, the agency stated, "Notwithstanding this required regulatory process, both President Obama and Administrator Jackson have repeatedly indicated their preference for comprehensive legislation to address this issue and create the framework for a clean energy economy."[19] A similar statement was made in the December press release announcing the finding's finalization. Nevertheless, in response to the *Massachusetts v. EPA* decision and other petitions and lawsuits, the agency is moving forward on several fronts to define how the Clean Air Act may be used.

Table 1. Petitions for Regulation of Greenhouse Gas Emissions Under the Clean Air Act

Date	Subject	CAA Section	Petitioner
10/20/99	New Motor Vehicles	202(a)(1)	International Center for Technology Assessment (ICTA) and 19 other organizations
10/3/07	Ocean Going Vessels	213(a)(4)	California Attorney General
10/3/07	Marine Shipping Vessels and their Fuels	213(a)(4) and 211	Oceana, Friends of the Earth, and the Center for Biological Diversity
1/10/08	New Marine Engines and Vessels	213(a)(4)	South Coast Air Quality Management District
12/5/07	Aircraft	231	States of California, Connecticut, New Jersey, New Mexico, Pennsylvania, City of New York, District of Columbia, South Coast Air Quality Management District
12/5/07	Aircraft Engines	231(a)(2)(A) and 231(a)(3)	Friends of the Earth, Oceana, the Center for Biological Diversity, and the Natural Resources Defense Council
1/29/08	New Nonroad Vehicles and Engines and Rebuilt Heavy-Duty Engines, excluding Aircraft and Vessels	202(a)(3)(D) and 213(a)(4)	ICTA, Center for Food Safety, and Friends of the Earth
1/29/08	New Nonroad Vehicles and Engines, excluding Aircraft, Locomotives, and Vessels	202 and 213(a)(4)	States of California, Connecticut, Massachusetts, New Jersey, Oregon, and Pennsylvania

Table 1. (Continued)

Date	Subject	CAA Section	Petitioner
7/29/09	Fuels Used in Motor Vehicles, Nonroad Vehicles, and Aircraft	211 and 231	NYU Law School Institute for Policy Integrity
9/21/09	Concentrated Animal Feeding Operations	111(b) and (d)	Humane Society of the United States and 8 other organizations

Source: U.S. EPA and the petitioning organizations.

GOING AFTER MOBILE SOURCES: TITLE II AUTHORITIES

Although there are a number of ways in which the Clean Air Act could be used to regulate greenhouse gas emissions, this chapter focuses on the authorities in Title II of the act, which deals with mobile sources of air pollution. As noted in our discussion of *Massachusetts v. EPA*, it was a petition to EPA that it control GHG emissions from new motor vehicles that precipitated much of the discussion of the agency's Clean Air Act authority to control GHGs, and brought the issue to the Supreme Court. The Court dealt with this petition by reversing EPA's denial of it and remanding it for the agency to decide whether the pollutants in question endanger public health or welfare.

Since then, the agency has received nine more petitions asking it to regulate GHGs, all but one focused on mobile sources and their fuels (see **Table 1**). In addition to new motor vehicles, these petitions cover aircraft, ocean-going ships and their fuels, motor fuels in general, and nonroad vehicles and engines – a category that includes construction equipment, farm equipment, logging equipment, outdoor power equipment, forklifts, marine vessels, recreational vehicles, and lawn and garden equipment.

The specifics of the Clean Air Act sections that give EPA authority to regulate pollution from these sources vary somewhat, but it is generally believed that an endangerment finding and decision to regulate GHGs in response to any one of the petitions could trigger regulation in response to them all. With that in mind, we look at the authorities provided under Title II for the various categories of mobile sources, and what EPA's use of these authorities for conventional pollutants indicates with regard to its ability to regulate greenhouse gases. (A separate report, CRS Report R40585, *Climate*

Change: Potential Regulation of Stationary Greenhouse Gas Sources Under the Clean Air Act, discusses the potential Clean Air Act regulatory tools for stationary sources.)

New Motor Vehicles

In general, Title II authorities, which address mobile sources of pollution, are better suited to the regulation of GHGs than most other parts of the act, because the language is written to provide a great deal of flexibility to the Administrator in setting emission standards. In Section 202 of the act, the Administrator is directed to "prescribe (and from time to time revise) ... standards applicable to the emission of any air pollutant from any class or classes of new motor vehicles or new motor vehicle engines, which in his judgment cause, or contribute to, air pollution which may reasonably be anticipated to endanger public health or welfare." The Administrator can (in fact, must) use this authority to issue standards for any air pollutant for which she makes an endangerment finding under Section 202.[20]

For the most part, the pollutants for which standards were set under Section 202 have been either criteria pollutants[21] (carbon monoxide and nitrogen oxides) or emissions that served as precursors to the formation of a criteria pollutant (ozone) through chemical reactions in the atmosphere.[22] In 1986, EPA added emission standards for particulate matter under Section 202; in 2004, the agency added a standard for formaldehyde, a carcinogen.

The Administrator is given substantial leeway in the design and implementation of motor vehicle regulations. The act states that the Administrator may establish categories for purposes of regulation based on "gross vehicle weight, horsepower, type of fuel used or other appropriate factors." In addition, the Administrator may delay the effective date of regulations as long as she finds necessary "to permit the development and application of the requisite technology, giving appropriate consideration to the cost of compliance within such period." Using this authority in regulating conventional pollutants, EPA has previously established different standards for cars than for SUVs, minivans, and light trucks; often, it has used weight or power classifications to set differing levels of emission standards, particularly for trucks; it has given manufacturers as much as four years lead time to develop emission controls; and it has set different standards based on the type of fuel an engine uses. Except for specific conventional pollutants mentioned in Section 202, the act does not specify a level of stringency (e.g., best available control technology) for prospective regulations.

In promulgating its current car and light truck standards, the agency set 11 different emission levels (some of them temporary), allowing the manufacturers to choose any of these "bins," provided that the company achieves a prescribed overall fleet average. Thus, there is precedent for applying varying requirements to different sized vehicles, for phasing in requirements, and for the use of fleet averages as opposed to a single standard applied to all models.

In the GHG standards for motor vehicles proposed in the September 28, 2009 *Federal Register*, EPA has used the concept of a vehicle's "footprint" to set differing standards for different size vehicles. As explained by EPA:

> These standards are based on CO_2 emissions-footprint curves, where each vehicle has a different CO_2 emissions compliance target depending on its footprint value (related to the size of the vehicle). Generally, the larger the vehicle footprint, the higher the corresponding vehicle CO_2 emissions target. As a result, the burden of compliance is distributed across all vehicles and all manufacturers. Manufacturers are not compelled to build light vehicles of any particular size or type, and each manufacturer will have its own standard which reflects the vehicles it chooses it (sic) produce.[23]

Manufacturers will also be allowed to average, bank, and trade emission credits if the standards are promulgated as proposed.

Although flexible in many respects, motor vehicle standards have often been used to force the development of new technology. In adopting technology-forcing regulations, EPA has generally followed the lead of California. Because of its more severe air pollution and its pioneering role in establishing motor vehicle emission control requirements in the 1960s, California is allowed to adopt standards more stringent than federal requirements. The state must apply for a waiver of federal preemption in order to enforce its more stringent standards, which EPA is to grant if the state meets certain criteria, primarily a showing that the standards are needed to meet "compelling and extraordinary conditions."

In the 1970s, California and the federal government used technology-forcing regulations to bring about the development and introduction of catalytic converters, reducing emissions from new vehicles as much as 90% in less than a decade. In recent years, California's Zero Emission Vehicle and Low Emission Vehicle (LEV2) requirements, though frequently modified, have stimulated the development of alternative fuel, hybrid, and electric vehicles and led to the development of the National Low Emission Vehicle program and EPA's current Tier 2 emission standards.

Emissions of GHGs from new motor vehicles are the subject of already promulgated California standards, for which the state requested that EPA issue a waiver of federal preemption in December 2005. These California standards will require gradual reductions of GHG emissions from new passenger vehicles until they are about 30% below the emissions of the 2002 fleet in 2016. Compliance will be determined by fleet averages, rather than by the emissions of individual vehicles, and the regulations provide additional flexibility, including averaging of emissions over a five-year period, banking, and trading of credits within and among manufacturers. (For a discussion of the California standards, see CRS Report RL34099, *California's Waiver Request Under the Clean Air Act to Control Greenhouse Gases From Motor Vehicles*.) The California program, for which EPA granted a waiver June 30, 2009, has served as a model for how EPA could regulate motor vehicle GHGs under the Clean Air Act.

Promulgation of an endangerment finding triggers the development of regulations for emissions of four greenhouse gases emitted by motor vehicles (CO_2, methane, nitrous oxide, and hydrofluorocarbons).[24] According to EPA, emissions of the four gases from motor vehicles (passenger cars, light duty trucks, other trucks and buses, and motorcycles) accounted for 23.6% of the total inventory of U.S GHG emissions in 2006. Most of the emissions are in the form of CO_2 (see Figure 1), which is the product of combusting any fuel containing carbon. Hydrofluorocarbons (HFCs), the chemicals used as coolants in auto air conditioning systems, are the second most important motor vehicle GHG; but, as the figure shows, they are a distant second.

In setting standards for emissions of motor vehicle GHGs, EPA can either regulate the GHGs as a basket, or regulate them individually. The agency has proposed a combination of the two approaches, with hard caps on emissions of methane and nitrous oxide, but averaging, banking, and trading of credits allowed for CO_2 and HFCs. The endangerment finding proposal stated that the Administrator believes she can regulate emissions of the GHGs either individually or as a group: "If the Administrator defines the air pollutant as the group of greenhouse gases, she believes she would have the discretion to set standards that either control the emissions of the group as a whole, and/or standards that control emissions of individual greenhouse gases, as constituents of the class."[25] This determination, like other elements of the proposal, was subject to public comment.

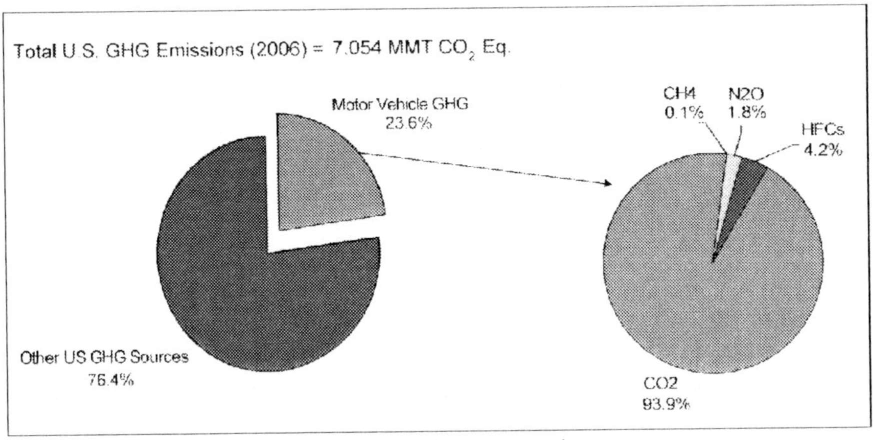

Source: U.S. EPA, March 6 Draft Deliberative Presentation

Notes: Motor vehicles = passenger cars, light duty trucks, other trucks, buses, and motorcycles, including releases of HFCs from motor vehicle air conditioning. CH4=methane; N2O=nitrous oxide; HFCs=hydrofluorocarbons.

Figure 1. Motor Vehicle Greenhouse Gas Emissions

As shown in Table 2, cars and light trucks (pickups, minivans, SUVs, etc.) account for three- quarters of motor vehicle GHG emissions, with medium- and heavy-duty trucks accounting for most of the rest. The regulations proposed on September 28 apply only to cars and light trucks, not to medium- and heavy-duty trucks.

Table 2. Motor Vehicle GHG Emissions, 2007, by Source Category (Million Metric Tonnes, CO_2-e)

Category	Total GHG Emissions	% of Motor Vehicle Total
Passenger Cars	664.5	40.3%
Light Duty Trucks	561.6	34.0%
Medium- and Heavy-Duty Trucks	410.4	24.9%
Buses	12.4	0.8%
Motorcycles	2.0	0.1%
Total	1650.9	

Source: U.S. EPA, *Draft Inventory of U.S. Greenhouse Gas Emissions and Sinks, 1990-2007*, Table 2-15.

Although they would affect a large group of emission sources that account for a significant percentage of total U.S. GHG emissions, the effectiveness of motor vehicle emission standards in reducing total GHG emissions is limited in that both California's and the proposed EPA standards will apply only to new motor vehicles. The auto and light truck fleet turns over slowly: the median survival rate for 1990 cars, for example, was 16.9 years, and that for light trucks was 15.5 years.[26] Given this durability, the impact of GHG standards on the total emissions of the motor vehicle fleet will take a long time to be felt. If historic experience is any guide, reductions in GHG emissions per new vehicle, unless they are very aggressive, may be largely offset by growth in vehicle miles traveled.

Ships

Three of the 10 petitions to EPA asking the agency to control greenhouse gas emissions concern ocean-going ships (also referred to as marine engines and vessels) and (in two of the petitions) their fuel. Although there is a wide range of estimates, the International Maritime Organization's consensus is that international shipping emitted 843 million metric tonnes of carbon dioxide, 2.7% of global CO2 emissions in 2007. Including domestic shipping and fishing vessels larger than 100 gross tonnes, the amount would increase to 1.019 billion tonnes, 3.3% of global emissions.[27] At these levels, only five countries (the United States, China, Russia, India, and Japan) account for a higher percentage of the world total of CO2 emissions.[28]

In addition to the CO_2 emissions, the low quality bunker fuel that ships use and the absence of pollution controls result in significant emissions of black carbon and nitrogen oxides, which also contribute to climate change. Refrigerants used on ships (hydrofluorcarbons and perfluorocarbons—HFCs and PFCs) are also potent greenhouse gases when released to the atmosphere. Thus, the total impact of ships on climate may be somewhat greater than 3%.

The authority to control pollution from ships is found in Section 213(a)(4) of the Clean Air Act, which provides general authority to the Administrator to promulgate standards for emissions other than carbon monoxide, oxides of nitrogen, and volatile organic compounds from "nonroad engines and vehicles."[29] Fuels are regulated separately under Section 211 of the act.

The language of Section 213 is similar to that for new motor vehicles in Section 202. If the Administrator determines that emissions of GHGs from ships "significantly contribute"[30] to air pollution which may reasonably be anticipated to endanger public health or welfare, she may promulgate such regulations as she deems "appropriate." Except for the specific conventional pollutants mentioned in Section 213(a)(2), there is no level of stringency (such as best available control technology) specified for prospective regulations. The Administrator may establish classes or categories of ships for the purposes of regulation. There is no deadline for the promulgation of standards, and in setting them, the Administrator may take into account costs, noise, safety, and energy factors associated with the application of technology.

A wide variety of measures might be undertaken to reduce emissions from shipping, from simple operational measures, such as reducing speed[31] or using cleaner fuels, to various hull and propeller design features that would increase fuel economy. Reducing speed can save substantial amounts of fuel. A.P. Moller-Maersk, which operates the world's largest fleet of containerships, reported that it reduced fuel consumption by its ships 6% in 2008, compared to the fuel used for the same level of business activity in 2007. According to the company, "Reducing speed 5-10% does increase the number of days at sea, but reduces both fuel consumption and CO_2 emission by more than 15%." The petitions also mention improved fleet deployment planning, use of shore- side power while in port, heat recovery systems, the use of sails as supplemental propulsion sources, and NOx controls, such as selective catalytic reduction (SCR) or exhaust gas recirculation, as potential emission control measures.

A complicating factor in the regulation of emissions from ocean-going vessels would be that, for the most part, their GHG emissions occur in international waters, and the sources (the ships) are not registered in the United States: according to California's petition, 95% of the fleet calling on U.S. ports is foreign-flagged. The petitioners assert that these factors are not a bar to EPA regulation, however, citing as precedent a Supreme Court case that held that the Americans with Disabilities Act could be applied to foreign-flagged cruise ships so long as the ADA-required accommodations did not interfere with the ships' internal affairs or require major, permanent modifications to the ships.[32] In addition, according to the California petition, the United States can and does enforce pollution standards on ships in its territorial waters, "as can be seen by the fact that the National Park Service has imposed air pollutant emissions controls on cruise ships, including foreign-flagged cruise ships (the vast majority of such ships are foreign-flagged), that sail off the coast from Glacier Bay National Park, in Alaska."[33]

EPA did not address these assertions in the ANPR, but it might be that a distinction needs to be made between operational factors (speed, fuel type, etc.), that are easily amenable to control, and permanent modifications of the vessel. In its petition, California cites the Clean Air Act's historic role in forcing the development of new, cleaner technology, but the Supreme Court opinions in *Spector v. Norwegian Cruiseline* and other cases might seem to impose some limits on the degree to which the act's authority can be used with respect to foreign-flagged vessels.[34]

In addition to petitioning for regulation of emissions from ships, the petitions from California and from Oceana et al. stated that EPA should regulate the composition of marine shipping vessel fuel to control global-climate-change-related emissions, or should require use of marine diesel fuel oil instead of bunker fuel. The purpose would be to limit the sulfur content of marine fuels and reduce NOx emissions. We discuss EPA's authority to regulate fuels in a separate section below, but note here that EPA, the State of California, and the International Maritime Organization are all moving forward with regulations to limit the sulfur content of bunker fuel for the purpose of reducing conventional pollutants. California's low sulfur fuel requirements went into effect July 1, 2009. In addition, on March 27, 2009, EPA proposed that the entire U.S. coastline except portions of Alaska be designated by the IMO as an Emission Control Area, subject to lower sulfur limits in bunker fuel.

Sulfur emissions form fine particles of sulfate in the atmosphere, with significant impacts on public health and welfare. (For a further discussion of these impacts, see CRS Report RL34548, *Air Pollution and Greenhouse Gas Emissions from Ships*.) Although harmful as a conventional pollutant, sulfur emissions are thought by most experts to be beneficial or at least neutral in the climate context. Sulfates have a cooling effect on the atmosphere, since the particles tend to reflect solar radiation back into space rather than absorbing it. On the other hand, removing sulfur might be necessary to prevent the fouling of pollution control equipment that reduces other pollutants that do lead to warming.

Other Nonroad Engines

Section 213 can also be used to regulate other nonroad vehicles and engines. A similar endangerment finding would first be required, following which the Administrator may promulgate such regulations as she deems appropriate to control emissions from the classes or categories of nonroad engines that she determines "significantly contribute to" the air pollution that endangers public health or welfare. The Administrator is to take into account costs, noise, safety, and energy factors in setting standards. There is no deadline for setting standards.

The nonroad sector is a broad category that includes construction equipment, farm equipment, forklifts, outdoor power equipment, lawn and garden equipment, and recreational vehicles.[35] This group accounted for 199.7 million metric tons of CO2 emissions in 2007, according to the two petitions requesting regulation (see Table 3), 3.3% of total U.S. emissions of CO2 in that year. According to the ICTA petition, GHG emissions from the nonroad sector increased 49% between 1990 and 2005, a higher rate of emissions increase over the same period than for on-road vehicles (32%), aircraft (3%), boats and ships (36%), and rail (32%).[36]

Given their smaller impact on overall emission levels, EPA has been slower to regulate conventional (criteria) pollutants from nonroad engines than from motor vehicles. Many of these engines had few emission control requirements for as many as 25 years after the regulation of automobiles. In the last decade, however, often following the lead of California, EPA has promulgated standards for many nonroad categories. Some of these standards, particularly for diesel-powered equipment and for lawn and garden equipment have been technology-forcing. Others, such as for snowmobiles, have been less so.

In general, given the wide variety of engine types and sizes and the configurations of the equipment itself, the agency has based its standards on a review of individual subcategories and the technologies available to reduce emissions from specific types of machinery or equipment, rather than applying one across the board standard. Presumably, any GHG standards for this sector would take the same approach.

Aircraft

EPA has also received petitions to regulate GHG emissions from aircraft and aircraft engines. In the United States, aircraft of all kinds are estimated to emit between 2.4% and 3.4% of the nation's total greenhouse gas emissions.[37] When other factors are considered, the impact of U.S. aviation on climate change is perhaps twice that size. These factors include the contribution of aircraft emissions to ozone formation, the water vapor and soot that aircraft emit, and the high altitude location of the bulk of aircraft emissions. (For additional information on aircraft GHG emissions, see CRS Report R40090, *Aviation and Climate Change*.)

Table 3. Nonroad Sector CO_2 Emissions, 2007, by Source Category (Million Metric Tonnes)

Category	CO_2 Emissions	% of Non-road Total
Construction and Mining Equipment	63.9	32.0%
Agricultural Equipment	39.6	19.8%
Industrial Equipment	27.8	13.9%
Lawn and Garden Equipment	23.8	11.9%
Commercial Equipment	16.4	8.2%
Pleasure Craft	15.8	7.9%
Recreational Equipment	9.4	4.7%
Logging Equipment	1.9	1.0%
Airport Equipment	1.0	0.5%
Railroad Equipment	0.2	0.1%
Total	199.7	

Source: ICTA et al., *Petition for Rulemaking Seeking the Regulation of Greenhouse Gas Emissions from Nonroad Vehicles and Engines*. According to the petition, the emissions data were compiled by the Western Environmental Law Center using EPA's nonroad emissions model.

As noted in Table 1, two December 2007 petitions requested that EPA address aircraft GHG emissions. Specifically, the petitions asked that EPA make a finding that aircraft GHG emissions endanger public health or welfare, and that the agency adopt regulations that allow a range of compliance approaches: these might include emission limits, operational practices, fees, a capand-trade system, minimizing engine idling time, employing single engine taxiing, or use of ground-side electricity measures to replace the use of fuel-burning auxiliary power units at airport gates.[38]

EPA has authority to regulate emissions from aircraft under Section 231 of the Clean Air Act. The language is similar to that for other mobile sources. It requires the Administrator to issue standards for the emission of any air pollutant from any class or classes of aircraft engines which, in her judgment, causes or contributes to air pollution which may reasonably be anticipated to endanger public health or welfare. The regulations are to take effect "after such period as the Administrator finds necessary ... to permit the development and application of the requisite technology, giving appropriate consideration to the cost of compliance.... "

Unlike ships, aircraft operating in the United States are generally registered here: EPA has cited data that foreign carriers accounted for only 3% of major carrier operations in the United States in 1999.[39] Thus, whether GHG regulations could be applied to foreign flag carriers might seem to pose less of an issue, at least in terms of whether any potential regulations would address the bulk of the sector's U.S. emissions. On the other hand, international air travel is extremely competitive, and issues of whether regulations can be imposed on foreign carriers have already been raised in the context of the European Union's adoption of cap-and-trade requirements for international aviation. U.S. airlines generally maintain that the imposition of requirements on foreign-flag airlines (i.e., themselves, in the European Union) violates international trade agreements. Their preference is that any controls be negotiated through the International Civil Aviation Organization (ICAO) and be applied equally to all carriers.

It is also worth noting that, compared to other mobile sources, EPA's Clean Air Act authority vis- à-vis aircraft and aircraft engines contains two principal differences. The first is that the Administrator must consult with the Administrator of the Federal Aviation Administration and the Secretary of Transportation in developing emission standards, and is not allowed to change the standards if doing so would significantly increase noise and adversely affect safety. The second, and perhaps more significant difference, is that EPA has rarely regulated emissions from aircraft without first negotiating international agreements through ICAO. ICAO's regulation of conventional pollutants from aircraft, unlike EPA's regulation of the same pollutants from motor vehicles, has consistently avoided forcing technology. The most recent standards for nitrogen oxides, for example, essentially ratified what the principal aircraft manufacturers had already achieved.[40]

Fuels

Fuel regulation, whether of bunker fuel, gasoline, or any other type of fuel, is authorized under Section 211 of the Clean Air Act. Section 211 gives the Administrator authority to control or prohibit the manufacture and sale of any fuel or fuel additive if she concludes that its emission products may endanger public health or welfare, or if they will impair to a significant degree the performance of emission control devices. As with the regulation of engines and vehicles, the Administrator is given substantial leeway in the design and implementation of fuel regulations and there is no deadline for their promulgation even after an endangerment finding is made.

GHG emissions from fuels have already been targeted for regulation by the State of California.[41] On March 5, 2009, the California Air Resources Board (CARB) proposed the California Low Carbon Fuel Standard, the goal of which is to reduce GHG emissions from transportation fuels per unit of energy 10% by 2020.[42] The regulations would address emissions from the production, transportation, and consumption of gasoline, diesel fuel, and their alternatives, including biofuels.

The regulations envision compliance both through the use of lower carbon fuels and through the development of more efficient, advanced-technology vehicles, such as plug-in hybrids, electric vehicles, and hydrogen fuel cells. A public comment period on the proposed standards began with their release on March 5. CARB hopes to finalize the regulations by 2010.

As has been the case with motor vehicles, California has often led the way in the development of cleaner conventional fuels through technology-forcing regulation, with U.S. EPA later adopting similar standards. Thus, many view the Low Carbon Fuel Standard as the prototype of another possible use of existing Clean Air Act authority to regulate greenhouse gas emissions nationally. On July 29, 2009, the Institute for Policy Integrity at NYU Law School petitioned EPA to establish a cap-and-trade system to limit greenhouse gas emissions from fuels used in motor vehicles, nonroad vehicles, and aircraft.

Regulation of fuels would be a way for CARB or EPA to obtain reductions from existing vehicles and engines. As noted earlier, the slow turnover of the vehicle fleet means that emission reductions from new vehicles would only gradually affect emission levels from the fleet as a whole. By requiring low carbon fuels, CARB and EPA could obtain GHG reductions from the entire fleet more quickly.

On the other hand, measuring the carbon content of fuels is more complicated than it may seem, particularly if one considers the life-cycle emissions, including indirect impacts of production. EPA is embroiled in a controversy over this issue already,[43] as it attempts to develop a methodology for measuring greenhouse gas emissions from biofuels, as required by the Energy Independence and Security Act of 2007 (P.L. 110-140).[44] For regulations implementing this provision, EPA is still developing a methodology to measure the GHG effects of indirect land-use changes, such as the switching of land from forest to cropland.

Table 4. Categories of Sources Whose GHG Emissions Could Be Regulated Under Title II of the Clean Air Act (Assuming an Endangerment Finding for the Category)

Category	CAA Authority (Section #)	Estimated 2007 GHG Emissions (million tonnes CO_2-e)	% of Total U.S. GHG Emissions
Passenger Cars	202	664.5	9.3%
Light Duty Trucks	202	561.6	7.9%
Medium- and Heavy-Duty Trucks	202	410.4	5.8%
Aircraft (domestic operation)	231	189.4	2.7%
Construction and Mining Equipment	213	63.9	0.9%
Agricultural Equipment	213	39.7	0.6%
Ships and Other Boats[a]	213	33.7	0.5%
Industrial Equipment	213	27.8	0.4%
Lawn and Garden Equipment	213	23.8	0.3%
Commercial Equipment	213	16.4	0.2%
Pleasure Craft	213	15.8	0.2%
Buses	202	12.4	0.2%
Recreational Equipment	213	9.4	0.1%
Motorcycles	202	2.0	<0.1%
Logging Equipment	213	1.9	<0.1%
Airport Equipment	213	1.0	<0.1%
Railroad Equipment	213	0.2	<0.1%
Total[b]		2073.9	29.1%

Source: U.S. EPA and ICTA

a. Does not include international bunker fuel.

b. Emission reductions from these categories might also be addressed through regulation of the carbon content of their fuels.

Summary of Mobile Source and Fuel GHG Emissions

Table 4 summarizes EPA's existing authorities over mobile source GHG emissions and the emissions of each of the sectors discussed in this chapter.

Given the Supreme Court's remand in *Massachusetts v. EPA*, the agency is focusing its efforts on motor vehicles (passenger cars and light trucks), which, as Table 4 shows, account for the majority of mobile source GHG emissions. Nevertheless, having issued an endangerment finding for motor vehicles, it would appear that EPA could move forward to control GHG emissions from most of the other categories of mobile sources and/or their fuels by issuing similar endangerment findings.

Regulation of GHGs from mobile sources might also lead the agency to establish controls for stationary sources, such as electric power plants.[45] That option, which would use authorities in different sections of the Clean Air Act than Title II, is addressed in CRS Report R40585, *Climate Change: Potential Regulation of Stationary Greenhouse Gas Sources Under the Clean Air Act*, by Larry Parker and James E. McCarthy. Electric power plants accounted for 33.6% of U.S. GHG emissions in 2007.

End Notes

[1] For a more detailed discussion of cap-and-trade approaches to GHG emission control, see CRS Report RL33799, *Climate Change: Design Approaches for a Greenhouse Gas Reduction Program*. For additional information on allowance allocation methods, see CRS Report RL34502, *Emission Allowance Allocation in a Cap-and-Trade Program: Options and Considerations*.

[2] Now also being called "cap-and-tax" by some opponents, or "cap-and-dividend" by some supporters.

[3] For a more detailed discussion of a carbon tax approach to GHG emission control, see CRS Report R40242, *Carbon Tax and Greenhouse Gas Control: Options and Considerations for Congress*.

[4] The reverse assumption, that a cap would *not* be adjusted even if it caused significant economic dislocation, is presumed by cap-and-trade proponents. Both assumptions would likely be tested.

[5] Memorandum from Jonathan Z. Cannon, EPA General Counsel, to Carol M. Browner, EPA Administrator, "EPA's Authority to Regulate Pollutants Emitted by Electric Power Generation Sources," April 10, 1998.

[6] The lead petitioner was the International Center for Technology Assessment (ICTA). The petition may be found on their website at http://www.icta.org/doc/ghgpet2.pdf.

[7] The agency argued that it lacked statutory authority to regulate greenhouse gases: Congress "was well aware of the global climate change issue" when it last comprehensively amended the Clean Air Act in 1990, according to the agency, but "it declined to adopt a proposed amendment establishing binding emissions limitations." Massachusetts v. EPA, 549 U.S. 497 (2007).

[8] Memorandum from Robert E. Fabricant, EPA General Counsel, to Marianne L. Horinko, EPA Acting Administrator, "EPA's Authority to Impose Mandatory Controls to Address Global Climate Change Under the Clean Air Act," August 28, 2003.

[9] Massachusetts v. EPA, 549 U.S. 497 (2007). The majority held: "The Clean Air Act's sweeping definition of 'air pollutant' includes '*any* air pollution agent or combination of such agents,

including *any* physical, chemical ... substance or matter which is emitted into or otherwise enters the ambient air.... ' ... Carbon dioxide, methane, nitrous oxide, and hydrofluorocarbons are without a doubt 'physical [and] chemical ... substances[s] which [are] emitted into ... the ambient air.' The statute is unambiguous."

[10] For further discussion of the Court's decision, see CRS Report RS22665, *The Supreme Court's Climate Change Decision: Massachusetts v. EPA*, by Robert Meltz.

[11] U.S. EPA, "Regulating Greenhouse Gas Emissions Under the Clean Air Act," 73 *Federal Register* 44354, July 30, 2008.

[12] Ibid., p. 44356.

[13] Ibid.

[14] In December 2007, EPA staff did reach a conclusion that greenhouse gas emissions contribute to air pollution that may reasonably be anticipated to endanger public welfare and prepared a draft endangerment finding stating so; but the Bush Administration chose not to release it. The draft Bush-era finding was eventually released to the public by the Obama Administration's EPA, October 13, 2009. See "Bush EPA Planned Endangerment Finding Based on Public Welfare, Document Shows," *Daily Environment Report*, October 15, 2009, p. A-1. The text of the Bush-era draft can be found at *http://www.epa.gov/foia/docs/2007* _Draft_Proposed_Endangerment_Finding.pdf.

[15] U.S. EPA, "Proposed Endangerment and Cause or Contribute Findings for Greenhouse Gases Under Section 202(a) of the Clean Air Act," 74 *Federal Register* 18886, April 24, 2009.

[16] The proposed regulation would require light-duty vehicles to meet an estimated combined average emissions level of 250 grams/mile of CO_2 in model year 2016. Under the CO2 fleet average standard proposed under the Clean Air Act, EPA expects manufacturers to take advantage of an option to generate CO_2-equivalent credits by reducing emissions of hydrofluorocarbons (HFCs) and CO_2 through improvements in their air conditioner systems, as well as by improving fuel economy. NHTSA is to set corresponding fuel economy standards, achieving a combined fuel economy of 34.1 miles per gallon for cars and light trucks by 2016, with intermediate standards beginning in the 2012 model year. 74 *Federal Register* 49468, September 28, 2009. For additional information, see CRS Report R40 166, *Automobile and Light Truck Fuel Economy: The CAFE Standards*, by Brent D. Yacobucci and Robert Bamberger.

[17] "Proposed Rulemaking To Establish Light-Duty Vehicle Greenhouse Gas Emission Standards and Corporate Average Fuel Economy Standards," 74 *Federal Register* 49454, September 28, 2009.

[18] Although generally referred to as simply "the endangerment finding," the EPA Administrator actually finalized two separate findings: a finding that six greenhouse gases endanger public health and welfare and a separate "cause or contribute" finding that the combined emissions of greenhouse gases from new motor vehicles and new motor vehicle engines contribute to the greenhouse gas pollution that endangers public health and welfare. The findings will appear in the *Federal Register*. A pre-publication copy can be found at http://www.epa.gov/climatechange/endangerment.html.

[19] "EPA Finds Greenhouse Gases Pose Threat to Public Health, Welfare / Proposed Finding Comes in Response to 2007 Supreme Court Ruling," Press Release, April 17, 2009, at http://yosemite.epa.gov/opa/admpress.nsf/d0cf661 8525a9efb85257359003fb69d/0ef7df675805295d8525759b00566924.

[20] There is no need to demonstrate that the regulations will eliminate the danger posed by the air pollution in question. In the case of other air pollutants—notably, ozone—progressively more stringent standards for motor vehicle emissions have been promulgated for more than 40 years without the pollutants the regulations addressed being reduced sufficiently for all nonattainment areas to attain the ambient air quality standards.

[21] Criteria pollutants are pollutants that endanger public health or welfare and whose presence in ambient air results from numerous or diverse sources. Sections 108 and 109 of the act give the Administrator the authority to identify such pollutants and set national ambient air

quality standards for them. There are six criteria pollutants, all of which were identified in the 1970s.

[22] Ozone is not generally emitted directly by emission sources, but forms in the atmosphere as the result of chemical reactions among precursors in the presence of sunlight. The precursors include the hydrocarbons present in gasoline (currently termed "non-methane organic gases," NMOG, in the regulations), as well as nitrogen oxides and carbon monoxide, which are regulated as criteria pollutants in their own right.

[23] The regulations are at 74 *Federal Register* 49454, September 28, 2009. For a further description, including emission targets for specific types of vehicle and specifics regarding the averaging, banking , and trading of emission credits, see "EPA and NHTSA Propose Historic National Program to Reduce Greenhouse Gases and Improve Fuel Economy for Cars and Trucks," Fact Sheet, at *http://www.epa.gov/otaq/climate/regulations* /420f09047.htm#2.

[24] Two other commonly mentioned greenhouse gases, sulfur hexafluoride (SF_6) and perfluorocarbons, are not emitted by motor vehicles.

[25] Environmental Protection Agency, "Proposed Endangerment and Cause or Contribute Findings for Greenhouse Gases Under Section 202(a) of the Clean Air Act," 74 *Federal Register* 18905, April 24, 2009.

[26] Oak Ridge National Laboratory, for the U.S. Department of Energy, *Transportation Energy Data Book: Edition 27*, 2008, Tables 3.10 and 3.11.

[27] International Maritime Organization, *Updated Study on Greenhouse Gas Emissions from Ships*, Executive Summary of Phase 1 Report, 1st September 2008, p. 5 at egserver.unfccc. int/seors/attachment/file_storage/6ep77qqvcujba7k. doc. Both estimates exclude emissions from naval vessels.

[28] Oceana, *Shipping Impacts on Climate: A Source with Solutions*, p. 2, at *http://www.ocean* a.org/fileadmin/oceana/uploads/Climate_Change/Oceana_ Shipping_Report.pdf.

[29] CO, NOx, and VOCs are regulated under Section 213(a)(3), which requires the imposition of best available control technology, and set a deadline for such regulation.

[30] Section 213 uses the words "significantly contribute," whereas Section 202 says "cause, or contribute to."

[31] See *Preparing for the Future*, The A.P. Moller – Maersk Group's Health, Safety, Security and Environment Report 2008, pp. 28-3 0, at http://media

[32] Spector v. Norwegian Cruiseline, 545 U.S. 119 (2005).

[33] People of the State of California Acting by and Through Attorney General Edmund G. Brown, Jr., "Petition for Rule Making Seeking the Regulation of Greenhouse Gas Emissions from Ocean-Going Vessels," October 3, 2007, p. 13. The cited regulations are at 36 CFR 13.65(b)(4). The *Federal Register* citation is 61 *Federal Register* 27008, 27011 (May 30, 1996).

[34] In addition to these considerations, U.S. international treaty obligations, specifically international trade agreements, may be implicated. These issues are beyond the scope of this chapter.

[35] Locomotives are also considered nonroad engines, but the authority to regulate them (in Section 213(a)(5)) is separate from that applicable to the rest of the category, and does not appear to give the Administrator the authority to set GHG standards. Thus, they were not included in either of the two petitions EPA received requesting regulation of nonroad GHG emissions.

[36] International Center for Technology Assessment, et al., "Petition for Rulemaking Seeking the Regulation of Greenhouse Gas Emissions from Nonroad Vehicles and Engines," January 29, 2008, p. 5.

[37] The lower percentage includes CO_2 emissions from consumption of fuel by military aircraft, general aviation, and domestic operation of commercial aircraft. The higher estimate includes CO_2 emissions from international air travel originating in the United States, as well.

[38] For a brief discussion of the petitions, see 73 *Federal Register* 44460, July 30, 2008. Some of these measures, such as minimizing engine idling time, employing single engine taxiing, and use of ground-side electricity measures to replace the use of fuel-burning auxiliary power units, are already widely used by the airlines as fuel-saving measures.

[39] U.S. EPA, Office of Air and Radiation, *Emission Standards and Test Procedures for Aircraft and Aircraft Engines, Summary and Analysis of Comments*, November 2005, p. 10, at http://www.epa.gov/oms/regs/nonroad/aviation/ 420r05004.pdf

[40] "EPA Proposal to Bring Certain Aircraft Up to International Engine Standard," *Daily Environment Report*, September 30, 2003.

[41] For more information, see CRS Report R40078, *A Low Carbon Fuel Standard: State and Federal Legislation and Regulations*.

[42] The proposed regulation and supporting information can be found on CARB's website at http://www.arb.ca.gov/ regact/2009/lcfs09/lcfsisor1 .pdf.

[43] For information, see CRS Report R40460, *Calculation of Lifecycle Greenhouse Gas Emissions for the Renewable Fuel Standard (RFS)*.

[44] Section 202 of the act mandates the use of "advanced biofuels"—fuels produced from non-corn feedstocks and with 50% lower lifecycle greenhouse gas emissions than petroleum fuel—starting in 2009. Of the 36 billion gallons of renewable fuel required in 2022, at least 21 billion gallons must be advanced biofuel.

[45] In its ANPR, in July 2008, the agency stated: "As part of EPA's efforts to respond to the Supreme Court's decision, the Agency conducted a thorough review of the CAA to identify and assess any other CAA provisions that might authorize regulation of GHG emission sources. That review made clear that a decision to control any source of GHG emissions could or would impact other CAA programs with potentially far-reaching implications for many industrial sectors. In particular, EPA recognized that regulation of GHG emissions from motor vehicles under section 202(a)(1) or from other sources of GHG emissions under many other provisions of the Act would subject major stationary sources to preconstruction permitting under the CAA." 73 *Federal Register* 44399, July 30, 2008.

In: Cars, Climate and the EPA: Issues in... ISBN: 978-1-61728-963-7
Editors: Carolynne D. Burgdoerfer

Chapter 2

LEGAL CONSEQUENCES OF EPA'S ENDANGERMENT FINDING FOR NEW MOTOR VEHICLE GREENHOUSE GAS EMISSIONS

Robert Meltz

SUMMARY

On December 15, 2009, the Environmental Protection Agency (EPA) took its most important action to date related to climate change. EPA published its final determination that the combined greenhouse gas (GHG) emissions from new, light-duty motor vehicles in the United States contribute to an "endangerment" from climate change. More precisely, EPA found that such emissions, in the words of Clean Air Act (CAA) section 202(a), "cause or contribute to air pollution that may reasonably be anticipated to endanger public health or welfare." Under section 202(a), this finding requires that EPA promulgate "standards" to control such emissions—as the agency proposed to do in advance of its endangerment determination.

Some groups have objected to the endangerment determination and the emission standards to follow, arguing they will trigger a "cascade" of unacceptable regulatory consequences under *other* CAA provisions. These regulatory consequences, they say, would impose unattainableGHG concentration goals on EPA and the states, and/or economically and administratively unreasonable burdens. This chapter examines the CAA provisions that have figured in this debate to see whether this alleged cascade of legal consequences likely would occur.

First, the report examines CAA sections that, like section 202(a), are triggered by endangerment findings. Of these, the one most likely to require EPA regulatory action after the 202(a) endangerment finding is section 111, authorizing new source performance standards—but only as to stationary source categories emitting the largest amounts of GHGs. Section 111, however, affords EPA wide discretion in setting new source performance standards. Two other sections that arguably might be triggered are 108, requiring national ambient air quality standards, and 115, which requires states to revise their implementation plans to prevent or eliminate the endangerment of public health or welfare in a foreign country. As to these sections, however, the arguable infeasibility of the regulatory goals—even if GHG emissions in the United States are significantly reduced, atmospheric concentrations would decline little—will give EPA room to argue that regulatory action is not mandatory. Other endangerment-triggered sections of the CAA can be distinguished from section 202(a) by their explicit terms, and thus would likely not be triggered by the 202(a) endangerment finding.

Second, the report looks at CAA provisions having no endangerment trigger. Of these, EPA has conceded that two require the agency to act after it promulgates the required emission standards following the 202(a) endangerment finding. One provision would require EPA to impose "best available control technology" (BACT) on GHG emissions from any major emitting facility proposed to be constructed in a Prevention of Significant Deterioration area. The other, Title V, creates an operating permit program for stationary sources of emissions, and would require stationary sources subject to BACT under the first provision to also apply for Title V permits. As to each of these requirements, EPA has proposed a "tailoring rule" setting emission thresholds far higher than in the CAA, at least for a few years. EPA justifies the departure from statutory language under the case law doctrines of "absurd results" and "administrative necessity."

A caveat: the issue analyzed in this chapter is important primarily if Congress does not enact climate change legislation that puts regulation of GHGs beyond the reach of some of the CAA provisions discussed here. In particular, the House-passed climate change bill, H.R. 2454 (the American Clean Energy and Security Act of 2009), states that three of the CAA sections treated in this chapter, and one CAA title, may not be used to address air pollutants based on their climate.

PART ONE: INTRODUCTION

On December 15, 2009, the Environmental Protection Agency (EPA) published its most important action to date related to climate change. It found that the combined greenhouse gas (GHG) emissions from new, light-duty motor vehicles in the United States contribute to an "endangerment" from climate change.[1] More precisely, EPA found that such emissions, in the words of Clean Air Act (CAA) section 202(a),[2] "cause or contribute to air pollution that may reasonably be anticipated to endanger public health or welfare."[3] Under section 202(a), this final determination requires that EPA promulgate "standards" to control such emissions from new light-duty vehicles.

It is not, however, the likelihood of standards for *new motor vehicle* GHG emissions that has sparked controversy. Indeed, EPA's proposed standards for emissions of carbon dioxide (CO_2, the principal GHG) from light vehicles were set so as to align with the Department of Transportation's simultaneously proposed fuel economy standards, and are reported to be achievable with presently available technology.[4] Rather, objection has been raised to the section 202 endangerment finding for GHGs and the upcoming motor vehicle GHG emission standards because of the argument they will trigger a cascade of unacceptable regulatory consequences under *other* CAA provisions. These regulatory consequences, objectors say, would impose unattainable air quality goals on EPA and the states, and/or economically and administratively unreasonable burdens on both EPA and regulated entities.

This chapter examines the CAA provisions that have figured in this debate to assess whether this alleged "cascade" of legal consequences would likely occur as the result of EPA's endangerment finding under 202(a) and resulting emission standards. Looked at another way, the report assesses the legal leeway EPA has should it wish to limit its expansion of GHG regulation beyond new motor vehicles under section 202(a).[5] The reader also may wish to consult EPA's own analysis of this issue in an Advance Notice of Proposed Rulemaking (ANPR) issued in 2008,[6] and the analyses submitted to EPA in response to the ANPR. The *policy* pros and cons of using CAA provisions other than 202(a) to address GHGs are treated in CRS Report R40585, *Climate*

Change: Potential Regulation of Stationary Greenhouse Gas Sources Under the Clean Air Act, by Larry Parker and James E. McCarthy.

A caveat: the issue analyzed here is important primarily if Congress does not enact climate change legislation that puts regulation of emissions based on their climate change impacts beyond the reach of some of the CAA provisions discussed here. In particular, the House-passed climate change bill, H.R. 2454, states that sections 108 (national ambient air quality standards) and 115 (international air pollution) may not be used to address air pollutants based on their climate change impacts. The bill also would prohibit the use of section 165 (requiring preconstruction "new source review" in Prevention of Significant Deterioration areas) to regulate GHG emissions, and specifies that no stationary source of GHGs could be required to obtain a permit under Title V of the CAA solely because of its GHG emissions.

Part Two of this chapter offers some history as to how EPA came to issue its endangerment determination for new, light-duty motor vehicle GHGs on December 15, 2009. Part Three assesses EPA's leeway, following its endangerment determination, in deciding whether to regulate GHG emissions from other sources—where those sources are covered by CAA provisions that, like section 202(a), have an "endangerment" finding as the trigger. Part Four gauges the effect of the same endangerment determination, or more accurately the standards that must follow, under CAA provisions that lack an endangerment determination trigger. Part Five offers a summary and comment.

Part Two: The Path to Now

The events leading up to EPA's endangerment finding for new light-duty motor vehicle GHGs in 2009 began a decade earlier. In 1999, 19 organizations and 13 states petitioned EPA to regulate the four GHGs emitted by new motor vehicles (carbon dioxide, methane, nitrous oxide, and hydrofluorocarbons). Petitioners invoked CAA section 202(a):

> The [EPA] Administrator shall by regulation prescribe ... standards applicable to the emission of any air pollutant from any class or classes of new motor vehicles ... which in his judgment cause, or contribute to, air pollution which may reasonably be anticipated to endanger public health or welfare. ....

Petitioners argued that GHGs are "air pollutants" under the term's definition in CAA section 302(g),[7] and that GHGs "endanger public health and welfare" because of their impact on global climate. Thus, petitioners concluded, EPA must regulate such emissions through standards under section 202(a).

EPA rejected the petition in 2003, concluding that GHGs are not "air pollutants" under the CAA so the agency lacked authority to regulate GHG emissions from new motor vehicles based on their climate change effects.[8] Moreover, said EPA, even if it did have the authority, it would be imprudent for the agency to exercise it at that time because of political opposition in Congress and various Bush Administration policies opposing mandatory controls on GHGs. EPA's rejection of the petition was appealed to the federal appeals court in Washington, D.C., which upheld the agency's decision 2-1.[9]

In 2007, the Supreme Court in *Massachusetts v. EPA* reversed the appeals court 5-4, finding for Massachusetts.[10] The Court held that for CAA purposes, GHGs are "air pollutants" and hence that EPA has the power under the act to regulate GHG emissions from new motor vehicles. Moreover, said the Court, the phrase "in his judgment" in section 202(a) does not permit EPA to avoid regulation of air pollutants such as GHGs based on policy considerations. That is, an EPA decision not to grant a section 202 petition must relate not to administration policy, but to whether an air pollutant causes, or contributes to, air pollution which may reasonably be anticipated to endanger public health or welfare.[11] That being so, the EPA can avoid regulating GHGs under 202(a) only if it finds that GHGs "do not contribute to climate change or if it provides some reasonable explanation as to why it cannot or will not exercise its discretion to determine whether they do."[12] The Supreme Court decision did not order EPA to make a positive endangerment finding, only to resolve the petition asking it to regulate vehicle GHGs on a correct understanding of the CAA.

In response to *Massachusetts v. EPA*, the EPA under the Bush Administration moved toward an endangerment determination in 2007 but ultimately backed off. On April 24, 2009, under the newly arrived Obama Administration, EPA issued a proposed positive endangerment determination for GHGs from new light-duty motor vehicles.[13] In response, it received 380,000 comments. On September 28, 2009, EPA also proposed the aforementioned *emission standards* for GHGs from new light-duty motor vehicles—the standards required by CAA section 202(a) and referred to in this chapter's title. [14] (September, 2009, was a busy month for EPA on the climate change front. Besides proposing vehicle emission standards, the agency went final with its GHG monitoring rule for large stationary sources,[15] and proposed a "tailoring rule" discussed in Part Four of this chapter.) Most recently, EPA published its final endangerment determination for GHGs from new light-duty motor vehicles on December 15, 2009.

Actually, EPA's endangerment determination consists of *two* findings, one addressing each component of the section 202(a) endangerment provision. First, EPA found that current atmospheric levels of the six major GHGs—from all emission sources—constitute in combination "air pollution which may reasonably be anticipated to endanger public health and welfare." This is a generic determination, not tied to any particular source category. Second, EPA found that emissions of GHGs by new motor vehicles (comprising four of the aforementioned six GHGs) will in combination "cause or contribute to" those atmospheric levels. In contrast to the generic determination, this finding is plainly linked to one source category. This bifurcation of the 202(a) endangerment language is important in Part Three of this chapter. In Part Three, a reference to a CAA endangerment provision being satisfied or simply to a "positive endangerment finding" means that *both* the "may reasonably be anticipated to endanger" component finding and the "contributes to" component finding are satisfied.

PART THREE: CAA SECTIONS WITH ENDANGERMENT TRIGGERS OTHER THAN SECTION 202(A)

To what extent will the endangerment finding under section 202(a) narrow EPA's discretion to deal, or not deal, with GHGs under other CAA provisions with endangerment preconditions? The CAA contains about a dozen such "endangerment" preconditions. Each one, like section 202(a), establishes a precondition for EPA action against an air pollutant: EPA first must find that emissions of that air pollutant (or group of pollutants) from a particular source category "cause or contribute to air pollution which may reasonably be anticipated to endanger public health or welfare." Some endangerment provisions, not section 202(a), add the qualifier "significantly" after "contribute," and there are slight variations in singular/plural and punctuation; otherwise the quoted language repeats throughout the statute. Depending on the CAA section, if EPA finds an endangerment, it is required or at least authorized, usually without other preconditions, to take action addressing the endangerment by the source category.

At the outset, an administrative law question arises. Since this chapter's topic is whether EPA might be forced to regulate GHG emissions under CAA provisions outside 202(a), one must ask: Does EPA have to respond *at all* to petitions it receives for endangerment findings and standard setting under those provisions? (At least 10 such petitions seeking EPA control of GHG emissions from this or that source category have been filed with the agency since *Massachusetts v. EPA*.[16]) Can the agency simply ignore requests to regulate GHG emissions, or inform a petitioner that EPA is studying the matter without ever reaching a decision? None of the CAA endangerment provisions expressly authorizes petitions to the agency, in contrast with other provisions in the CAA and elsewhere that do establish petition mechanisms with agency deadlines for response.[17]

The answer is that while courts afford federal agencies wide discretion as to whether to initiate a rulemaking,[18] the discretion is not without limit. The Administrative Procedure Act is the touchstone: under section 553(e), each agency must give interested persons "the right to petition for the issuance ... of a rule,"[19] which petition is subject to section 555(b)'s command that an agency "within a reasonable time ... proceed to conclude a matter presented to it."[20] If an agency fails to heed these directives, section 706(1) empowers courts to "compel agency action unlawfully withheld or unreasonably delayed."[21] The Supreme Court recently explained that an APA section 706(1) claim can proceed only when an agency failed to take a discrete agency action that it is required to take[22]—but petitions seeking EPA regulation of GHG emissions would presumably argue that this is exactly what they are alleging. Moreover, these APA provisions have been applied not only where an agency responds to a rulemaking petition by denying it, but to instances such as those hypothesized here, where the agency refuses to respond at all.[23] In sum, a reviewing court likely would require EPA to respond—up or down—to a well-supported petition for GHG regulation within a generous, but finite, period of time.

Given that EPA responds to the petition (or acts without petition), one issue that arises under every non-202(a) endangerment provision is the effect of the *generic* component of the agency's 202(a) finding. Recall that the generic component declared that GHGs already in the atmosphere from all sources may, in combination, reasonably be anticipated to endanger public health or welfare. Because non-202(a) endangerment provisions use identical language, arguably at least the generic component of those provisions also must be deemed satisfied. This is an important preliminary step in understanding the legal relevance of EPA's 202(a) endangerment finding to other endangerment-triggered sections of the CAA—but it is only a preliminary step. Beyond this, as the following discussion shows, there are differences in the language surrounding the endangerment provisions, and lower levels of GHG emissions from most non-section-202 source categories. These differences, it would appear, generally give EPA leeway in deciding whether to regulate GHGs from sources other than new motor vehicles, notwithstanding the 202(a) finding.

Following are the CAA endangerment provisions being debated as possibly authorizing, or even mandating, EPA regulation of GHG emissions now that EPA has issued its 202(a) endangerment determination.

Fuels and Fuel Additives

CAA section 211(c)(1)[24] empowers EPA to regulate the manufacture, sale, or offering for sale, of fuels and fuel additives for use in motor vehicles or nonroad engines or vehicles. The authority comes into being when, in EPA's judgment, any emission product of such fuel or fuel additive triggers an endangerment provision identical to that in 202(a). Unlike section 202, which authorizes standards for emissions, section 211 speaks of standards for the composition of the fuel that produces the emissions. Presumably because section 202 specifies only a final result, rather than dictating how manufacturers are to achieve that result, it is "the preferred—although not the mandatory—alternative under the statutory scheme."[25]

Section 211(c)(1) arguably allows EPA to implement low-carbon fuel standards.[26] However, section 211 (c)(1) states only that EPA "may" regulate if an endangerment finding is made— unlike section 202(a), which directs that EPA "shall" regulate on such a finding.[27] Assuming that "may" is used in the traditional sense of allowing full discretion, a 21 1(c)(1) endangerment finding would not compel agency action against fuels and fuel additives. That being so, it is unnecessary to address whether a "contributes to" component finding under 211 (c)(1) would be compelled by a comparable finding under 202(a).

National Ambient Air Quality Standards

CAA section 108(a)(1)[28] contains endangerment language identical to that of 202(a) and might provide a basis for regulating stationary sources of GHGs through national ambient air quality standards (NAAQSs).

By way of background, NAAQSs are nationwide standards set by EPA that fix the maximum concentration of a pollutant (or group of pollutants) allowed in ambient air.[29] There are two types of NAAQSs: primary NAAQSs to protect public health, and secondary NAAQSs to protect public welfare.[30] Under the CAA, the path to establishing primary and secondary NAAQSs for an air pollutant consists of three steps: (1) listing of the pollutant by EPA, which requires (2) issuance by the agency of "air quality criteria" (documents summarizing the scientific information about a pollutant's effects—not, as "criteria" implies, standards), which, in turn, requires (3) promulgation of the NAAQSs. States then are required to submit "implementation plans" to EPA within three years to achieve and/or maintain the NAAQSs. [31] As a practical matter, any concern that EPA may be forced by the 202(a) endangerment finding to promulgate NAAQSs arises chiefly from primary, not secondary, NAAQSs. The reason: areas designated nonattainment for a primary NAAQS must attain the NAAQS within five years from such designation; in contrast, no fixed deadline is set for attainment of secondary NAAQSs.[32]

As the above shows, listing the pollutant is the critical first step—listing compels criteria, which compel NAAQSs. Section 108(a)(1) states that "for the purpose of establishing NAAQSs," the EPA Administrator "shall" publish a list of air pollutants (A) which "cause or contribute to air pollution which may reasonably be anticipated to endanger public health or welfare"; (B) which are emitted from "numerous or diverse mobile *or stationary* sources"; *and* (C) "for which air quality criteria had not been issued before the date of enactment of the Clean Air amendments of 1970, but for which he plans to issue air quality criteria under this section."[33] The endangerment finding under section 202(a) arguably forces EPA at some point to find condition (A) satisfied— the precise phrase "cause or contribute to air pollution which may reasonably be anticipated to endanger public health or welfare" is found in both 108(a)(1) and 202(a), and 108(a)(1)'s reference to "contribute to air pollution" presumably includes the same new motor vehicle emissions EPA just found to do so under 202(a). As for condition (B), that CO2 is emitted by "numerous and diverse mobile or stationary sources" seems hard to deny. But based on the literal language, EPA could be expected to argue that condition

(C) gives it discretion whether to "plan[] to issue air quality criteria"—hence discretion not to promulgate a NAAQS.[34]

There is contrary authority denying such discretion, however. In *NRDC v. Train*,[35] the Second Circuit held that EPA could *not* use condition (C) to avoid setting a NAAQS for lead, once the agency had recognized lead as satisfying conditions (A) and (B). To hold otherwise, said the court, would render the 108(a)(1) language that EPA "shall" publish a list "mere surplusage," and allow EPA to bypass the rigid CAA deadlines for attaining air quality standards.[36] Moreover, said the court, "it is to the initial list alone [the one required to be published soon after enactment of the 1970 CAA amendments] that the phrase 'but for which he plans to issue air quality criteria' is directed."[37] Plainly, *Train* might be invoked by any party seeking to compel EPA to set NAAQSs for GHGs following a positive endangerment finding for GHGs under section 202(a).

But *Train* does not settle the matter. Most significantly, the consequences of listing GHGs under section 108(a), leading to NAAQSs for GHGs, are so awkward—some would say absurd—that an argument arises that Congress could not have intended this result (assuming NAAQSs for GHGs were set below current atmospheric concentrations[38]). For one thing, the state implementation plans (SIPs) that would be required to achieve these NAAQSs would have an impossible goal, since even a substantial reduction of GHG emissions in an air quality control region likely would have little effect on ambient concentrations of GHGs. For this reason, a court might well take seriously an argument that an endangerment finding under section 202(a) does not force EPA to promulgate NAAQSs for GHGs. Courts tend to resist interpretations of statutes that lead to impractical or absurd results.[39]

Other arguments undermining the present-day viability of *NRDC v. Train* can be conceived, though each is debatable. First, *Train* dealt specifically with lead, which the Senate committee report for the 1970 CAA explicitly expected would be the subject of a NAAQS.[40] The Second Circuit twice noted this explicit mention of lead in the Senate report. In contrast, there is no legislative history indicating a congressional expectation of a NAAQS for GHGs, suggesting that *Train* would have less force in this context.

Second, EPA notes in the ANPR that *NRDC v. Train* was decided before the Supreme Court's decision in *Chevron U.S.A. Inc. v. NRDC*.[41] *Chevron* sets out the canonical statement of a key judicial review principle: courts should defer to an agency's interpretation of a statute it is charged with administering.[42] According to *Chevron*, if the statute is ambiguous or silent on a point, a court is to ask only whether the agency's interpretation is a "permissible" construction thereof.[43] The reviewing court is not to substitute its own judgment for a permissible agency construction, even if the agency's interpretation is not the only reasonable one. EPA's argument in the ANPR, it appears, is that the Second Circuit might have deferred to EPA's we-have-discretion argument had the court written its decision post-*Chevron*. However, this argument overlooks the fact that the deference principle was well established long before *Chevron*, and certainly by the time of *NRDC v. Train*.[44]

Finally, one might ask why, if Congress always intended NAAQSs to be required once conditions (A) and (B) are met, no new NAAQSs for a new criteria pollutant have been set since the 1970s.[45]

New or Modified Stationary Sources

CAA section 111 (b)(1) commands EPA to issue new source performance standards (NSPSs) for new or modified stationary sources of air pollution (principally, factories and power plants) once it determines that a stationary source category "causes, or contributes significantly to, air pollution which may reasonably be anticipated to endanger public health or welfare." Immediately plain is one reason why a 202(a) endangerment finding may not force EPA to promulgate a GHG NSPS for most new source categories: the term "significantly" in 11 1(b)(1) distinguishes that provision from 202(a). "Significantly" is not defined in the CAA, so EPA likely has broad discretion to define it to exclude many stationary source categories.

EPA's discussion accompanying its 202(a) endangerment finding, however, extensively documents the harm GHGs may have on the environment, and notes that even GHG source categories that appear too small to matter in fact could be very significant contributors in terms of both absolute emissions or in comparison to other similar source categories. Such statements may limit EPA's ability to define "significantly" so as to eliminate entirely any obligation to deal with GHGs through NSPSs, especially for the higher emitting source categories. In particular, coal- fired power plants emitted 27.6% of total U.S. GHG emissions in 2008[46]—by most any measure a "significant" contribution. Thus, the generic component of EPA's 202(a) endangerment finding, in combination with a new source category that is a "significant" contributor of GHGs, would seem to trigger an EPA duty to promulgate NSPSs for that source category.

In the case of section 111, however, the bare duty to promulgate an NSPS for a new source category's GHGs is the minor part of the story. More important is the considerable discretion EPA would have as to the precise content and applicability of that NSPS. For this reason, the agency is likely to favor NSPSs for dealing with GHGs—that is, as long as it has to address GHGs under the existing CAA. The discretion inheres first in the definition of NSPS, which directs EPA to factor in the feasibility of reducing emissions with current technology, and the cost.[47] In the statute's words, NSPSs must reflect the degree of emission reduction achievable through—

> the best system of emission reduction which (taking into account the costs of achieving such reduction and any nonair quality health and environmental impact and energy requirements), the [EPA] Administrator determines has been adequately demonstrated.[48]

The D.C. Circuit has interpreted this standard to mean that EPA should "identify the emission levels that are 'achievable' with 'adequately demonstrated technology'" (an undefined phrase allowing EPA broad discretion) and then use its discretion "to choose an achievable emission level which represents the best balance of economic, environmental, and energy considerations."[49] Also, section 111(b)(2) allows EPA to "distinguish among classes, types, and sizes within categories of new sources" in setting NSPSs, adding yet another layer of agency discretion. In sum, even if the 202(a) endangerment finding leads to an NSPS being required for GHGs emitted by some new source category, EPA's latitude in formulating that standard makes it unlikely the standard will be overly burdensome.

At present, there are several active suits seeking to compel EPA promulgation of NSPSs for GHG emissions from one or another stationary source category.[50] Petitioners' prospects in such suits likely will be improved by EPA's 202(a) endangerment finding.

Non-202(a) Mobile Sources of Air Pollution

Non-road Engines and Non-road Vehicles

CAA section 213[51] applies to non-road engines and non-road vehicles, which include ocean-going vessels, construction equipment, farm tractors, forklifts, and lawn and garden equipment. Under 213(a)(4), if EPA determines that any emissions from new non-road engines or vehicles "significantly contribute to air pollution which may reasonably be anticipated to endanger public health or welfare, the Administrator may promulgate ... such regulations as the Administrator deems appropriate ... taking into account costs, noise, safety, and energy factors." This contrasts in three ways with section 202(a), which (1) requires, rather than merely allows,[52] EPA to issue regulations for (2) emissions that merely "contribute," rather than "significantly contribute," to air pollution, and (3) lacks the discretion-enhancing phrase "as the Administrator deems appropriate." Thus EPA has great discretion in responding (or not responding) to GHG emissions from the non-road category.

Aircraft Engines

CAA section 231[53] authorizes EPA to establish emissions standards for aircraft engines. The agency "shall, from time to time" propose such standards, once a 202(a)-identical endangerment provision is satisfied. Plainly, a "contributes to" finding for new motor vehicles under 202(a) has limited relevance to whether EPA must make a counterpart finding for a different source category, like aircraft, that emits far less GHGs. The pivotal question is whether a "contributes to" component finding may have to be made for aircraft anyway, independent of section 202(a), in that whatever the quantity of GHG emissions from aircraft, it is not insignificant. A "contributes to" component finding for aircraft GHGs combined with a "may reasonably be anticipated to endanger" component finding carried over from section 202(a) will force EPA to make an overall endangerment finding for aircraft. This argument may turn on the breadth of EPA's authority to set a de minimis threshold for the quantity of aircraft GHG emissions that is deemed to "contribute to."

What if an endangerment finding *is* made for aircraft GHG emissions? Unlike with 202(a), such a finding under 231 does not automatically require regulations. Section 231(a)(2)(B)(ii) states that EPA "shall not change the aircraft engine emissions standards" if doing so would "significantly increase noise and adversely affect safety."[54] Section 231(c) states further that no regulation shall apply if disapproved by the President following the Secretary of Transportation's finding that EPA regulations would create a hazard to aircraft safety.[55] Also, "any regulation prescribed under [section 231] ... shall take effect after such period as the Administrator finds necessary ... to permit the development and application of the requisite technology, giving appropriate consideration to the cost of compliance within such period."[56] Finally, even should standards be compelled, section 231 is silent as to their form and stringency level (as under section 202(a)).

Noting these features, the D.C. Circuit has held that EPA enjoys broad discretion under section 231 in determining appropriate regulations for aircraft emissions in light of cost, technology, and safety, and can subordinate emissions reduction to these other concerns.[57]

Air Pollutants from the Loading and Unloading of Tank Vessels

CAA section 183(f)(1)(A)[58] requires EPA to regulate "the emission of [volatile organic compounds] and any other air pollutant from the loading and unloading of tank vessels"[59] that EPA finds "causes, or contributes to, air pollution that may be reasonably anticipated to endanger public health or welfare."[60] Volatile organic compounds are not GHGs, but the phrase "any other air pollutant" presumably includes methane, which is a GHG and may be released during the loading and unloading of tank vessels. The same point made above under section 231 applies here. To reiterate, a key question is whether a "may reasonably be anticipated to endanger" component finding under 202(a) will force EPA to make an endangerment finding for loading and unloading tank vessels given the argument that, however low relative to motor vehicles the GHG emissions from such operations, they nonetheless demand a "contributes to" component finding. The answer to this question, to reiterate, may turn on EPA's discretion in setting a de minimis threshold for the loading and unloading of tank vessels. Query whether the absence of the phrase "in his judgment" in 183(f)(1)(A), if contrasted with the presence of the same phrase in section 202(a), may be read by a court to constrain the agency's discretion under 183(f)(1)(A).

Even when promulgation of standards is compelled, EPA has discretion in other regards: section 183(f)(1)(A) does not explicitly specify the form and stringency of the required standards, and the standards are to reflect broad parameters (reasonably available technology, environmental impacts, energy requirements, and safety), as to which latitude of interpretation is inevitable.

Stratospheric Ozone Protection

CAA section 615[61] chiefly seeks to protect stratospheric *ozone*, but its broad language extends to *all* threats to the stratosphere. The section is included here because there are several scenarios under which GHGs might affect the stratosphere.[62]

The wording of section 615 is a bit different from the foregoing endangerment provisions:

> If, in the Administrator's judgment, any substance, process, practice, or activity may reasonably be anticipated to affect the stratosphere, especially ozone in the stratosphere, and such effect may reasonably be anticipated to endanger public health or welfare, the Administrator shall promptly promulgate regulations respecting the control of such substance, process, practice, or activity....

This language presents several distinctions from section 202(a), indicating that a positive endangerment finding under 202(a) would almost certainly not be found to compel EPA action under section 615. In particular, section 615 speaks of an *effect on the stratosphere* that may endanger public health or welfare, not, as does section 202(a), "air pollution" in the atmosphere generally that may do so. Thus, the generic component finding under section 202(a) does not dictate the comparable component finding here. For another thing, section 615 demands that the "substance" may *itself* be expected to affect the stratosphere in a way that endangers public health or welfare; section 202(a) sets a lower threshold, insisting only that the emissions "contribute to" existing air pollution levels that, without regard to the contributing source, may be expected to endanger. And as with some other endangerment provisions discussed in this memorandum, section 615, should it be deemed to require regulations, says nothing explicit as to their form or stringency.

Pollution in a Foreign Country Caused by Emission Sources in the United States

Under CAA section 115, "[w]henever the [EPA] Administrator, upon receipt of reports ... from any duly constituted international agency has reason to believe that any air pollutant or pollutants emitted in the United States cause or contribute to air pollution which may reasonably be anticipated to endanger public health or welfare *in a foreign country*," it must notify the governors of the states where the emission originates.[63] The same result, EPA notification of states, is required when the U.S. Secretary of State "alleges" such an endangerment and requests EPA to so notify. The notified states must then revise their state implementation plans "to prevent or eliminate" the endangerment being caused in the foreign country, but only if EPA finds that the same rights are granted to the United States by the country affected.[64]

Unlike other CAA provisions, section 115's endangerment provision imposes no requirement that EPA determine whether the air pollutant endangers public health or welfare *in the United States*. At first blush, therefore, one might conclude that a section 202(a) endangerment finding, based on effects in the United States, has no relevance to section 115, based on foreign effects. However, given the near uniformity of worldwide atmospheric concentrations of GHGs,[65] a positive 202(a) finding for the United States, which necessarily includes a "may reasonably be anticipated to endanger" finding for the United States, seems to compel the same component finding for any foreign country. As for the "contributes to" component finding, the question arises whether such finding is to be made for the United States overall, or for individual states. The former seems the better reading of section 115, given that the section speaks of pollutants emitted "in the United States" and refers to supporting studies by a "duly constituted international agency." In either event, EPA, upon making an endangerment finding and reciprocity finding respecting a foreign country, would be required to notify and demand implementation plan revisions from many, probably most, states.

As with setting NAAQSs for GHGs, the consequences of a section 115 endangerment finding (assuming reciprocity) are extremely awkward, and for the same reason. Again, the compelled EPA action would be a demand that many or even most state implementation plans be revised "to prevent or eliminate" the foreign endangerment. As discussed under the NAAQSs section of this chapter, such an ambitious result is likely unattainable, even assuming substantial reductions of GHG emissions in every plan-revising state, owing to the staggering number of GHG sources worldwide that contribute to the endangerment in the foreign country. Thus, with section 115 as with the CAA's NAAQS sections, EPA may search for a plausible interpretation of the statutory language that could avoid such consequences and the courts may be receptive.

Several such interpretations may be noted, though it must be stressed that they are of varying degrees of legal persuasiveness and have mostly not been tested in court. The first, noted by EPA in its ANPR, is that section 115 can only be used to require states to incorporate into their implementation plans *measures related to attainment and maintenance of NAAQSs.*[66] EPA's argument appears to be founded on section 115's statement that following an endangerment finding, EPA's notice to a state "shall be deemed to be a finding under section 110(a)(2)(H)(ii) ... which requires a plan revision with respect to so much of the applicable implementation plan as is inadequate to prevent or eliminate the endangerment ...."[67] Implementation plans are required to be prepared by states under section 110 to achieve and maintain NAAQSs,[68] and 110(a)(2)(H)(ii) states one of the required elements in such plans. Thus, EPA apparently concludes, section 115, being dependent on 11 0(a)(2)(H)(ii) for effectuation, likewise must be confined to measures related to attainment and maintenance of NAAQSs. If this argument is upheld, EPA might be able to avoid using section 115 if it can avoid setting a NAAQS for GHGs under section 110.

The vulnerability of EPA's argument, however, is that contrary to EPA's position, the text of section 110(a)(2)(H)(ii) appears to reach beyond measures related to attainment and maintenance of NAAQSs. Section 110(a)(2)(H)(ii) explicitly requires that implementation plans provide for plan revision when EPA finds that "the plan is substantially inadequate to attain the [NAAQS] which it implements *or to otherwise comply with any additional requirements established under this Act*."[69] Section 115's command that an EPA-notified state revise its plan to eliminate the endangerment would seem to be precisely such an "additional requirement[]." Moreover, that requirement applies under 115(a) to "any" air pollutant emitted in the United States, not just those that affect attainment and maintenance of NAAQSs.

A corollary to EPA's argument that 110(a)(2)(H)(ii) is limited to NAAQS-related measures is its stance that owing to the unavailability for GHGs of state notification and plan revision under 110(a)(2)(H)(ii), it is inappropriate for EPA to make the endangerment finding that sets that process in motion. This argument stems from the EPA view, endorsed by the D.C. Circuit, that section 115 establishes a single, unitary procedure.[70] That is, either both endangerment finding and notification/plan-revision occur, or neither does.

Still other arguments that would thwart judicial coercion of EPA action under section 115 are (1) that in contrast with the CAA section on inter*state* pollution, section 115 says nothing about allowing petitions and so they should be assumed unavailable,[71] and (2) that the text of section 115 ("Whenever" the Administrator "has reason to believe") implies a degree of discretion as to the endangerment finding.[72]

Part Four: CAA Sections Having No Endangerment Trigger

New Source Review in PSD Areas

Possibly the strongest arguments that a positive endangerment finding for new motor vehicle GHG emissions, and resulting standards, will compel EPA to act against other GHG sources is based on new source review in Prevention of Significant Deterioration (PSD) areas and on Title V permitting (discussed in the following section). EPA fully accepts this argument:

> EPA expects soon to promulgate regulations under the CAA to control GHG emissions [from new motor vehicles] and, *as a result, trigger PSD and Title V applicability requirements for GHG emissions.*[73]

As background, an area of the country is PSD for a pollutant if that area is either attaining the health-based NAAQS for that pollutant or there is insufficient information to determine whether the area is attaining that standard ("unclassifiable areas").[74] The PSD portion of the CAA seeks to ensure that air quality in such areas—air that is or may be "cleaner than clean"—not be allowed to deteriorate down to the minimum acceptable level in the NAAQSs. To ensure this does not happen, the PSD program acts preventively: it requires EPA or states with approved PSD programs to review in advance of construction any new major emitting facilities, or major modifications of existing facilities, proposed for the PSD area.[75] This is termed "new source review." As part of this review, the owner or operator of the proposed facility must show, for example, that emissions from the facility will not cause or contribute to more than one exceedance per year of the maximum allowable increase in concentration for any pollutant.[76] "Major emitting facility" is statutorily defined as any facility on a long statutory list (including large fossil-fuel-fired power plants) that has the potential to emit 100 tons/year of any air pollutant, and any stationary source in an unlisted category with the potential to emit 250 tons/year of any air pollutant.[77]

Since the PSD program is triggered by an area's being cleaner than a NAAQS requires, one might suppose it has no relevance to any pollutant for which there is no NAAQS—such as, so far, GHGs. But the CAA's PSD provisions cast a wider net. To be sure, an area can be *designated* PSD only in connection with a NAAQS air pollutant. But once so designated, the statute stipulates that "the proposed facility is subject to the best available control technology *for each pollutant subject to regulation under this Act* emitted from ... such facility."[78] The Supreme Court has held that GHGs are "air pollutants" under the CAA, but what does "subject to regulation under this act" mean? Did CO_2 become "subject to regulation" when EPA promulgated regulations in 1993 to require monitoring and reporting of CO_2 emissions,[79] as required by section 821 of the 1990 CAA amendments? Or later, in 2007, when the Supreme Court held that GHGs are "air pollutants"? Or still later, only when EPA promulgates an emission standard actually regulating CO_2 emissions from some source—such as the expected section 202 standard? In short, what event triggers the requirement of best available control technology (BACT) in PSD areas?

Much ink has been spilled over the meaning of "subject to regulation under this Act." Indeed, the phrase has been the topic of at least two EPA Environmental Appeals Board decisions and one EPA Administrator memorandum.[80] While the appeals board decisions have remanded the question to EPA regional offices for further consideration, the Administrator's memorandum adopts the narrowest possible reading of "subject to regulation." According to the memorandum, issued at the end of the Bush Administration, the phrase excludes pollutants for which EPA regulations only require monitoring and reporting, and includes only pollutants covered by an "actual control of emissions" of the pollutant under the CAA or its regulations. In 2009, EPA granted reconsideration of the Administrator's memorandum and has recently sought comment on the memorandum's interpretation and other interpretations of "subject to regulation."[81] But the dispute is likely to be moot soon, at least as to GHGs. Even under the narrow EPA interpretation, the agency's expected GHG standard for new motor vehicles in early 2010 will make certain GHGs "subject to regulation under this Act," and so trigger PSD/BACT for significant new sources of those pollutants.

A PSD/BACT requirement for CO2 raises a problem, however, owing to the low 100/250 tons/year thresholds for applicability of the PSD program—recall the definition of "major emitting facility" in the second paragraph of this section. Carbon dioxide is emitted in far greater amounts than any other air pollutant, so based on the statutory 250 tons/year threshold, hospitals, churches, schools, and other relatively minor sources may be covered. Besides the considerable pollution-control costs for small entities, coverage of such a vast number of sources, with a case-by-case determination of BACT for each source,[82] also raises questions of administrative practicality for state permitting authorities. To alleviate these problems, EPA has recently proposed a "tailoring rule" to tailor the tons/year thresholds of PSD new source review to federal and state administrative realities—in particular, to raise them substantially and thereby greatly reduce the number of covered sources.[83] The rule would depart, at least temporarily, from the explicit 100 or 250 tons/year threshold in the CAA. As described by EPA—

> The first phase, which would last 6 years, would establish a temporary level for the PSD and Title V applicability thresholds at 25,000 tons per year

> (tpy) on a "carbon dioxide equivalent" (CO_2e) basis, and a temporary PSD significance level for GHG emissions of between 10,000 and 25,000 tpy CO_2e. EPA would also take other streamlining actions during this time. ... EPA would conduct another rulemaking, to be completed by the end of the sixth year, that would promulgate, as the second phase, revised applicability and significance level thresholds and other streamlining techniques ....[84]

Of course, an agency's departure from the express and unequivocal language of a statute, as EPA proposes here, is something courts do not condone lightly—though it is another question entirely whether any group would *want* to file suit against the tailoring rule, and who would have standing to sue.[85] In any event, EPA goes to some effort in the preamble to the proposed tailoring rule to provide legal justification—citing two related, but independent, lines of analysis it terms the doctrines of "absurd results" and "administrative necessity." These lines of analysis apply both to P SD/new source review and Title V permitting, and so are discussed in a separate section of this chapter following the upcoming section on Title V.

BACT is defined in the CAA, but in a manner that leaves the permitting authority (EPA or the state) wide latitude in applying the requirement to a specific emission source.[86] That latitude is less, however, if EPA has an NSPS for the pollutant in question, since BACT can never result in emissions greater than allowed under an NSPS, if one exists.[87]

Title V Permitting

CAA title V, added by amendment in 1990, creates an operating permit program applicable to every "major" stationary source (existing and new), and certain other stationary sources, of air pollution.[88] However, the Title V permit program imposes no substantive requirements of its own. Rather, the purpose of a Title V permit is to consolidate in one place the requirements imposed on the source *elsewhere* in the CAA, thus simplifying the regulation of each source and the enforcement of such regulations by EPA, states, and citizen suits.

While imposing no new emission controls, Title V does impose costs on the applicant. For example, as part of the application process, covered facilities must identify all federally enforceable emission regulations (plus monitoring and recordkeeping requirements), including those in state implementation plans, and develop a compliance plan. Once the permit is received, the facility must monitor compliance and report the results to federal and state authorities at least every six months. Certain changes in facility operations require a formal Title V permit modification, at times a burdensome process. Finally, payment of permit fees is required, such fees to be not less than $25 per ton of regulated pollutant, with exceptions.[89] On the other hand, the act does seek to ameliorate these costs. For example, Title V authorizes states to issue "general permits" covering "numerous similar sources."[90] Facilities can then submit a simple application seeking coverage under the general permit.

The general threshold for coverage by Title V is that the facility directly emits, or has the potential to emit, 100 tons per year or more of any air pollutant. Such a low threshold would extend Title V permitting requirements to many stationary sources of GHGs that in addition fall under CAA emission requirements. In particular, a BACT requirement for GHG emissions in PSD areas, discussed above under New Source Review in PSD Areas, would require new major emitting facilities and major modifications in such areas to seek Title V permits.

As noted in the preceding section, the arguments made by EPA as to absurd results and administrative necessity are invoked as well by the agency for narrowing the application of Title V permitting. However, if the only CAA restrictions on stationary source GHGs are those under PSD new source review, the thresholds under that program will completely set the scope of Title V permitting.

"Absurd Results" and "Administrative Necessity"

EPA has sought to legally justify its departure from the tons/year thresholds stated by the CAA for P SD/new source review and Title V permitting by what it terms the "doctrines" of "absurd results" and "administrative necessity." These doctrines are discussed in detail in the preamble to the proposed tailoring rule, noted in the New Source Review in PSD Areas section of this chapter.[91]

The "absurd results" doctrine, supported by many court decisions, holds that even clear and explicit statutory directives may on rare occasion have to be departed from when rigid adherence would be inconsistent with other provisions of the statute or unambiguous legislative history, or produce a result so contrary to sensible policy as to permit the inference that Congress could not have intended that result. EPA asserts in the preamble to the proposed tailoring rule that adherence to the 100/250 tons/year threshold for new source review in PSD areas would have precisely this result—

> the extraordinary increases in PSD and Title V permit applications that would result from a literal application of the 100/250 tpy threshold requirements would, at least during the near term—until EPA and the permitting authorities can develop streamlining methods and ramp up resources—extensively disrupt the two permitting programs and impose undue regulatory burdens in the aggregate on the sources newly subject to PSD and Title V permit requirements.

As a concrete example, EPA asserts that during the initial period of literally applying the 100/250 tons/year threshold, permit applications would increase by 150-fold, "render[ing] it impossible for permitting authorities to meet the requirement in CAA section 165(c) to process permit applications within 12 months."[92]

From EPA's early years comes another example of courts endorsing an EPA effort to avoid a seemingly senseless consequence of a literal statutory reading. During the 1 970s, EPA adopted the position that the National Environmental Policy Act (NEPA) does not require EPA to do environmental impact statements, despite the fact that the act unequivocally makes its impact-statement directive applicable to "*all* agencies of the Federal Government."[93] The courts agreed with EPA, affirming the agency's rationale that it would be senseless to do environmental impact statements per NEPA command when the decision-making processes followed by EPA under its program statutes incorporate the very same environmental considerations and public participation as NEPA.[94] This may be regarded as an early and successful use of something akin to an absurd results principle.

In sum, if courts agree with EPA that the burdens imposed by a literal reading of the PSD and Title V thresholds would be crushing, it is quite plausible they would accept an EPA "absurd results" defense in any litigation attacking the tailoring rule. This is particularly so because EPA is not proposing a permanent exemption from the PSD and Title V thresholds, but rather a temporary phase-in period (though courts may want to revisit the EPA threshold when the agency makes its six-year adjustment in the threshold, as it promises to do). On the other hand, it must be noted that in the sheer magnitude of its consequences, the statutory departure in the proposed tailoring rule is greater than that in any of the "absurd results" cases with which CRS is familiar.[95]

The second, independent legal justification offered by EPA for its tailoring rule is the "doctrine" of "administrative necessity." This principle also has some support in the case law. As its name suggests, it authorizes an agency to depart from statutory commands when literal compliance is administratively impossible, or nearly so. To a considerable extent, administrative necessity and absurd results overlap. EPA strenuously argues in the proposed tailoring rule's preamble that the statutory PSD threshold (together with a "zero" significance level threshold for modifications to existing sources) "would result in a volume of permit applications that is so high that the PSD ... program[] would become impossible for State and Federal authorities to administer." Only if EPA can convince a court of such daunting burdens will the tailoring rule survive judicial review (again, *if* review is sought).

In the leading case on administrative necessity, *Alabama Power Co. v. Costle*,[96] the D.C. Circuit was also dealing with EPA, and, by coincidence, the PSD requirements of the CAA. While rejecting EPA's expansion of a statutory exemption, the court conceded that—

> limited grounds for the creation of exemptions are inherent in the administrative process, and their unavailability under a statutory scheme should not be presumed save in the face of the most unambiguous demonstration of congressional intent to foreclose them.[97]

The court stressed, unsurprisingly, that "[c]ategorical exemptions from the clear commands of a regulatory statute, though sometimes permitted, are not favored."[98] Yet, it added, "[t]his broad principle that frowns upon categorical administrative exemptions is strict, but not absolute."[99] It is possible that here, where EPA would have to apply the PSD program to a pollutant (CO2) emitted in far larger quantities than the pollutants for which the CAA was designed, EPA may be able to meet the demanding judicial prerequisites for invoking administrative necessity.

Finally, EPA has proposed that its phased approach to reaching the statutory thresholds may be justified by case law for the proposition that agencies may proceed in an incremental fashion in implementing a statutory mandate. EPA concedes that these cases are grounded on facts that are distinguishable from the tailoring rule situation, and solicits comment as to whether the case law is truly supportive of the proposed tailoring rule.[100]

PART FIVE: FINAL THOUGHTS

EPA's endangerment finding for new light-duty motor vehicle GHG emissions, or the emission standards that must follow, appears to force certain regulatory actions by EPA, leave the agency free to resist certain other regulatory actions if it chooses, and leave two additional regulatory authorities in a gray area in between.

Into the first category—forced regulatory action—appears to fall new source review of major emitting facilities and major modifications of existing facilities in PSD areas. Final motor vehicle GHG standards, expected by March, 2010, will necessarily mean that GHGs from new stationary sources that emit more than some threshold amount of any pollutant (100 or 250 tons/year under the CAA; 25,000 tons/year if the proposed tailoring rule is finalized), and GHGs from major modifications of existing facilities, will be subject to BACT in such areas. BACT, however, is loosely defined, particularly if there is no NSPS for the pollutant to set a floor on BACT. Title V permitting requirements likely will be triggered as well. Further, stationary source categories that emit large quantities of GHGs may have to be regulated by NSPSs, though the CAA gives EPA wide latitude in shaping such regulations.

The second category—EPA retains discretionary latitude—appears to include almost all the endangerment-triggered action authorities under the CAA. These can be distinguished in varying degrees from section 202(a) on the grounds that, unlike section 202(a), some of them use the discretionary "may," some demand that the source category contribute "significantly" to the endangerment, and some add additional preconditions. Most of these non-202(a) CAA authorities involve source categories whose GHG emission levels are lower than from new motor vehicles, meaning that a "contribute to" finding under section 202(a) may not dictate the same finding under that other authority. Stationary source categories that do not emit large amounts of GHGs will not have to be covered by NSPSs if EPA defines "significantly" at a high tons/year level.

The third category—the gray area between EPA being forced to act and not being forced to do so—appears to comprise promulgation of NAAQSs for GHGs under section 108 and international pollution abatement under section 115. These sections can be read to force EPA action, but there would appear to be wiggle room for agency interpretation given the awkward consequences and unattainable pollution-control goals such a reading could entail. The administrative necessity and absurd results arguments presented so vigorously by EPA in connection with PSD new source review and Title V permitting may be available to the agency here as well.

Ultimately whether EPA acts against GHG emissions under CAA provisions in the second and third categories is likely to be a function of the agency's (and Administration's) political desires more than anything else. The EPA under the Obama Administration has repeatedly stated that it is willing to take on GHG emissions through CAA regulation if Congress does not soon enact comprehensive climate change legislation reducing such emissions. There are several possible motivations for the agency's saying this: CAA regulation by the agency in the near future (1) ensures that time is not lost in abating the causes of climate change while Congress debates; (2) buttresses our credibility in international climate change negotiations (such as Copenhagen) by showing that the United States is "serious" about tackling climate change; and (3) puts pressure on Congress to enact climate change legislation.

End Notes

[1] 74 Fed. Reg. 66,496 (December 15, 2009). EPA announced its determination and posted it on its website a week earlier, on December 7, 2009, but it is publication in the Federal Register that starts the 30-day period before the determination takes effect.

[2] 42 U.S.C. § 7521(a).

[3] The term "welfare" is defined by the CAA to include, but not be limited to, effects on "soils, water, crops, vegetation, manmade materials, animals, wildlife, weather, visibility, and climate, damage to or deterioration of property, and hazards to transportation, as well as effects on economic values and on personal comfort and well-being...." 42 U.S.C. § 7602(h).

[4] *See* Proposed Rulemaking to Establish Light-Duty Vehicle Greenhouse Gas Emission Standards and Corporate Average Fuel Economy Standards, 74 Fed. Reg. 49,454 (September 28, 2009) (proposed jointly by EPA and the Department of Transportation's National Highway Traffic Safety Administration).

[5] *See, e.g., EPA chief signals opposition to Clean Air Act curbs on GHGs*, in GREENWIRE (December 8, 2009). (reporting the EPA Administrator's position that the agency opposes the use of national ambient air quality standards to reduce atmospheric concentrations of GHGs).

[6] 73 Fed. Reg. 44,354 (July 30, 2008). EPA concludes in the ANPR that "[a]n endangerment finding for GHGs under one provision of the Act could thus have ramifications under other sections of the Act." *Id.* at 44,418.

[7] The term "air pollutant" means any air pollution agent or combination of such agents, including any physical, chemical, biological, radioactive (including source material, special nuclear material, and byproduct material) substance or matter which is emitted into or otherwise enters the ambient air. Such term includes any precursors to the formation of any air pollutant, to the extent the Administrator has identified such precursor or precursors for the particular purpose for which the term "air pollutant" is used. CAA § 302(g); 42 U.S.C. § 7602(g).

[8] 68 Fed. Reg. 52,922 (September 8, 2003).
[9] 415 F.3d 50 (D.C. Cir. 2005).
[10] 549 U.S. 497 (2007).
[11] *Id.* at 532-53 3 (quotations marks and brackets omitted).
[12] *Id.* at 533.
[13] 74 Fed. Reg. 18,886 (April 24, 2009).
[14] 74 Fed. Reg. 49,454 (September 28, 2009).
[15] Though announced on September 22, 2009, this rule was not published until many weeks later: 74 Fed. Reg. 56,260 (October 30, 2009).
[16] A summary of seven of these petitions is in the ANPR mentioned on page 1 of this chapter. 73 Fed. Reg. at 44,458- 44,462. Of the seven petitions, three address GHGs from ocean-going vessels, two from aircraft, and two from nonroad engines and vehicles. On July 29, 2009, the Institute for Policy Integrity at the New York University School of Law submitted a petition addressing fuels used in motor and nonroad vehicles and engines, and aircraft. On September 21, 2009, a coalition of environmental, animal welfare, and environmental justice groups submitted a petition requesting EPA to promulgate new source performance standards for several air pollutants, including greenhouse gases, from new concentrated animal feeding operations. Finally, on December 2, 2009, the Center for Biological Diversity and *350.org* submitted a petition requesting EPA to issue national ambient air quality standards for seven GHGs.
Each of these petitions asks EPA to make the requisite endangerment finding under the appropriate CAA section— section 108, 111, 211, 213, or 231. These provisions are discussed further on in this part of the report. To date, EPA has not responded to any of the petitions. Finally, EPA has received petitions under section 115, dealing with endangerments in Canada owing to emissions in the United States. Because EPA does not recognize a petition process under section 115, it has characterized these self-described petitions as "submittals." The petitions/submittals under section 115 of which CRS is aware have been rejected.
[17] *See, e.g.*, CAA § 126(b), 42 U.S.C. § 7426(b). The text paragraph restates a point made by Justice Scalia in his four- justice dissent in *Massachusetts*. As he put it: "Where does the CAA say that the EPA Administrator is required to [make a 'judgment' under 202(a)] whenever a rulemaking petition is filed? The Court points to no such provision because none exists." 549 U.S. at 550. The majority does not address the question, probably because EPA did in fact render a formal decision on the petition in *Massachusetts*, denying it.
[18] As a leading commentator observes: "... courts have traditionally applied a highly deferential standard of review in reviewing cases of agency inaction." Jeffrey S. Lubbers, A GUIDE TO FEDERAL AGENCY RULEMAKING 542 (4th ed. 2006).
[19] 5 U.S.C. § 553(e).
[20] *Id.* at § 555(b).
[21] *Id.* at § 706(1).
[22] Norton v. Southern Utah Wilderness Alliance, 542 U.S. 55, 63 (2004).
[23] *See, e.g.*, In re: American Rivers and Idaho Rivers United, 372 F.3d 413, 418 (D.C. Cir. 2004).
[24] 42 U.S.C. § 7545(c)(1).
[25] Ethyl Corp. v. EPA, 541 F.2d 1, 11 (D.C. Cir. 1976) (en banc).
[26] Jameel Alsalam, Ctr. For Clean Air Policy, *A Pragmatic Approach to Regulating Greenhouse Gases Under the Clean Air Act* 14 (2009), *http://www.ccap.org/docs/resources* Clean%20Air%20Act%20and%20GHGs_CCAP_March%202009.pdf.
[27] Generally, unless the context indicates otherwise, the words "shall" and "may" in statutes are to be read in their everyday sense of mandatory and permissive, respectively. *See, e.g.*, Rastelli v. Warden, Metro. Correctional Center, 782 F.2d 17, 23 (2d Cir. 1986) ("The use of a permissive verb—'may review' instead of 'shall review'—suggests a discretionary rather than mandatory review process.").
[28] 42 U.S.C. § 7408(a)(1).

[29] *Id.* at § 7409(a). "Ambient air" is "that portion of the atmosphere external to buildings, to which the general public has access." 40 C.F.R. § 50.1(e).
[30] *See* footnote 3 *supra*.
[31] 42 U.S.C. § 7410(a)(1).
[32] *Id.* at § 7502(a)(2)(A)-(B).
[33] Emphasis added.
[34] *See* ANPR, 73 Fed. Reg. at 44, 477.
[35] 545 F.2d 320 (2d Cir. 1976).
[36] *Id.* at 325.
[37] *Id.* at 325.
[38] In particular, CAA section 109(b)(2) requires the "secondary" NAAQS to be set at a level that "protect[s] the public welfare from *any* known or anticipated adverse effects." (Emphasis added.) Since current atmospheric levels of GHGs are already believed to be producing welfare effects, it would seem that the plain meaning of the CAA language requires that a secondary NAAQS for GHGs be set lower than current concentrations. Under *Whitman v. American Trucking Ass 'n*, 531 U.S. 457 (2001), EPA may not consider costs of compliance in setting a NAAQS.
[39] *See, e.g.*, United States v. Turkette, 452 U.S. 576, 580 (1981) (in construing a statute, "absurd results are to be avoided"). In *Public Citizen v. U.S. Dep 't of Justice*, 491 U.S. 440, 470 (1989), Justice Kennedy, writing in concurrence for himself and two other justices, noted that "[w]here the plain language of the statute would lead to patently absurd consequences, that Congress could not *possibly* have intended, we need not apply the language in such a fashion. When used in a proper manner, this narrow exception to our normal rule of statutory construction does not intrude upon the lawmaking powers of Congress, but rather demonstrates a respect for the coequal Legislative Branch, which we assume would not act in an absurd way." (citations omitted; emphasis in original)
[40] S Rep. No. 91-1196, at 11 (1970).
[41] 467 U.S. 837 (1984). *See* ANPR, 73 Fed. Reg. at 44,477 n.229.
[42] Such deference has been held to be particularly appropriate in the case of "a statutory scheme as unwieldy and science-driven as the Clean Air Act." Appalachian Power Co. v. EPA, 135 F.3d 791, 801-802 (D.C. Cir. 1998).
[43] 467 U.S. at 843.
[44] Indeed, in the same year as *NRDC v. Train*, the Supreme Court explicitly deferred to EPA's interpretation of another one of its statutes, the Clean Water Act. *EPA v. California*, 426 U.S. 200, 227 (1976).
[45] The wording of the text sentence—"no new NAAQSs for a new criteria pollutant"—is meant to exclude revisions of existing NAAQSs (see CAA § 109(d)(1)), the replacement of the "total suspended particulates" NAAQS with a PM10 NAAQS in 1987, and the promulgation of the $PM_{2.5}$ NAAQS as a companion to the PM_{10} NAAQS in 1997.
[46] U.S. Energy Information Admin. (Department of Energy), EMISSIONS OF GREENHOUSE GASES IN THE UNITED STATES 2008 (2009). The 27.6% figure in the text was calculated by CRS from 2008 figures in Table 1 of the report, showing total U.S. GHG emissions as 5,839.3 million metric tons of CO_2 equivalent, and Table 11, showing coal-fired power plant emissions as 1,945.9 million metric tons of CO_2.
[47] 42 U.S.C. §§ 741 1(a)(1) and (h).
[48] *Id.* at § 7411(a)(1).
[49] Sierra Club v. Costle, 657 F.2d 298, 330 (D.C. Cir 1981).
[50] *See generally* CRS Report RL32764, *Climate Change Litigation: A Survey*, by Robert Meltz.
[51] 42 U.S.C. § 7547.
[52] *See supra* footnote 27 and accompanying text.
[53] 42 U.S.C. § 7571.
[54] *Id.* at § 7571(a)(2)(B).
[55] *Id.* at § 7571(3)(c).

[56] *Id.* at § 7571(2)(b).

[57] National Association of Clean Air Agencies v. EPA, 489 F.3d 1221, 1230 (D.C. Cir. 2007).

[58] 42 U.S.C. § 7511b(f)(1)(A).

[59] CAA section 183(f)(1)(A) incorporates the definition of "tank vessel" in 46 U.S.C. § 2101(39). A tank vessel is a vessel that carries hazardous material or oil in bulk, or is adapted to do so, and is (A) a vessel of the United States, (B) operates in U.S. waters, or (C) transfers its good in a U.S. port or any other place subject to U.S. jurisdiction.

[60] This endangerment provision lacks the "in his judgment" phrase found in CAA section 202(a).

[61] 42 U.S.C. § 7671n.

[62] *See* CRS Report R40585, *Climate Change: Potential Regulation of Stationary Greenhouse Gas Sources Under the Clean Air Act, supra* page 1, at 14.

[63] 42 U.S.C. § 74 15(a) (emphasis added). A comparable provision for abatement of *water* pollution in the United States causing endangerment in a foreign country is found at Clean Water Act section 310, 33 U.S.C. § 1320.

[64] *See* Peter Tsirigotis, *Discussion of Clean Air Act Authorities and GHGs*, available at http://www.epa.gov/air/caaac/ pdfs/2008_01tsirigotis.pdf (Jan. 31, 2008).

[65] With respect to the principal GHG, CO_2, see Japan Meteorological Society, *Carbon dioxide (CO_2) distribution* (updated May 20, 2009), available at *http://ds.data.jma.go.jp/ghg* /kanshi/info_kanshi_e.html.

[66] 73 Fed. Reg. at 44,482-44,483. *See also* Letter from Brian J. McLean, Director of EPA Office of Atmospheric Programs, to Albert Koehl, Ecojustice Canada, dated February 29, 2008 (rejecting petition for section 115 endangerment finding based on, among other U.S.-emitted pollutants, CO_2).

[67] 42 U.S.C. § 7410(a)(2)(H)(ii). The complementary CAA provision is section 1 10(k)(5): "Whenever the Administrator finds that the applicable implementation plan for an area is substantially inadequate to attain or maintain the relevant [NAAQS] ... *or to otherwise comply with any requirement of this Act*, the Administrator shall require the state to revise the plan as necessary to correct such inadequacies." (Emphasis added.)

[68] *Id.* at § 110(a)(1).

[69] Emphasis added.

[70] Her Majesty the Queen in Right of Ontario v. U.S. EPA, 912 F.2d 1525 (D.C. Cir. 1990).

[71] EPA makes this argument in the McLean letter, *supra* footnote 66, where the agency pointedly referred to the section 115 petition as a "submittal." The argument, however, seems to overlook the Administrative Procedure Act sections noted earlier in this chapter. *See* text accompanying notes 19-21 *supra*. Presumably EPA thinks of submittals as not triggering a duty to respond.

[72] *Ontario*, 912 F.2d at 1533.

[73] Prevention of Significant Deterioration and Title V Greenhouse Gas Tailoring Rule, 74 Fed. Reg. 55,292, 55,292 (October 27, 2009) (proposed rule). Emphasis added.

[74] CAA § 161; 42 U.S.C. § 7471.

[75] CAA§ 165; 42 U.S.C.§ 7475.

[76] CAA § 165(a)(3); 42 U.S.C. § 7475(a)(3).

[77] CAA§ 169(1); 42 U.S.C. § 7479(1).

[78] CAA § 165(a)(4); 42 U.S.C. § 7475(a)(4). Emphasis added. *See also* CAA § 169(3); 42 U.S.C. § 7479(3).

[79] 40 C.F.R. Part 75.

[80] Memorandum from Stephen L. Johnson, EPA Administrator, to EPA Regional Administrators, regarding "EPA's Interpretation of Regulations that [sic] Determine Pollutants Covered By Federal Prevention of Significant Deterioration (PSD) Permit Program," Dec. 18, 2008.

[81] 74 Fed. Reg. 51,535 (October 7, 2009). EPA identifies the current interpretation of "subject to regulation," from the Bush Administration, as the "current and preferred interpretation." 74 Fed. Reg. at 51,535.

[82] The CAA requires that BACT for a source be determined by the permit-issuing authority (the state or EPA) on a "case-by-case basis." CAA § 169(3); 42 U.S.C. § 7479(3).

[83] Prevention of Significant Deterioration and Title V Greenhouse Tailoring Rule, 74 Fed. Reg. 55,292 (October 27, 2009).

[84] *Id.* at 55,292. The threshold of 25,000 tpy relates solely to new major emitting facilities. "PSD significance level," on the other hand, relates solely to *modifications* of existing sources. As noted, only those modifications that are "major" trigger EPA new source review. (The qualifier "major" is not in the CAA; it was added by EPA regulation—see 40 C.F.R. § 52.21(a)(2)(ii)). To be major, a modification must increase the emissions from a source by at least an EPA- defined "PSD significance level." 40 C.F.R. §§ 52.21(b)(2)(i), 52.21(b)(40), 52.21(b)(23).

[85] It is likely that the additional GHGs emitted by virtue of the temporarily raised tons/year threshold in the tailoring rule would be insufficient to constitute the "injury in fact" required for standing to sue in federal courts. Speculating, however, it may be imagined that an environmental group might claim standing by arguing that its members were injured by the *non-GHG* emissions that would not have occurred had the statutory 100/250 tons/year threshold been applied. This argument is based on the fact that BACT for GHGs will also, at many sources, reduce non-GHGs as well. A second group that might have standing are industries that emit more than the 25,000 tons/year threshold in the tailoring rule, who could conceivably argue they are at an economic disadvantage to their competitors below the 25,000 tons/year threshold in the rule.

[86] CAA § 169(3); 42 U.S.C. § 7479(3).

[87] *Id.*

[88] CAA §§ 501-507; 42 U.S.C. §§ 7661-7661f.

[89] CAA § 502(b)(3)(B)(i); 42 U.S.C. § 766 1a(b)(3)(B)(i). As pertinent to GHGs, "regulated pollutant" is defined to include each pollutant regulated under section 111 (NSPSs) and each pollutant for which a primary NAAQS has been promulgated.

[90] CAA § 504(d); 42 U.S.C. § 7661c(d).

[91] 74 Fed. Reg. at 53,303-53,311 (absurd results doctrine); 74 Fed. Reg. at 53,311-53,318 (administrative necessity doctrine).

[92] 42 U.S.C. § 7475(c).

[93] 42 U.S.C. § 4332(2) (emphasis added).

[94] *See, e.g.*, State of Wyoming v. Hathaway, 525 F.2d 66, 71-72 (10th Cir. 1975). The question of NEPA's application to EPA was resolved, to some extent, by amendments to the Clean Air Act and Clean Water Act expressly exempting the agency from NEPA's impact statement requirement.

[95] Some industry groups have argued that EPA should defer promulgating section 202 GHG emissions standards so as to postpone triggering PSD/BACT for GHG emissions. This, of course, would avoid the problem of unrealistic statutory thresholds entirely.

[96] 636 F.2d 323 (D.C. Cir. 1979).

[97] *Id.* at 357.

[98] *Id.* at 358.

[99] *Id.* at 358.

[100] 74 Fed. Reg. at 55,319.

In: Cars, Climate and the EPA: Issues in... ISBN: 978-1-61728-963-7
Editors: Carolynne D. Burgdoerfer

Chapter 3

WAIVER AUTHORITY UNDER THE RENEWABLE FUEL STANDARD (RFS)

Brent D. Yacobucci

SUMMARY

Transportation fuels are required by federal law to contain a minimum amount of renewable fuel each year. This renewable fuel standard (RFS), established by the Energy Policy Act of 2005 (EPAct, P.L. 109-58) and amended by the Energy Independence and Security Act of 2007 (EISA, P.L. 110-140), requires that 12.95 billion gallons of renewable fuels be blended into gasoline and other transportation fuels in 2010. Most of this mandate will be met using corn-based ethanol. However, within the overall RFS there are secondary mandates for the use of cellulosic biofuels, biomass-based diesel fuels, and other advanced biofuels. Questions have been raised over whether there is enough feedstock supply and production capacity to meet these carveouts.

The Environmental Protection Agency (EPA) has the authority to waive the RFS requirements, in whole or in part, if certain conditions outlined in the law are present. In 2008 the governor of Texas requested a waiver of the RFS because of high grain prices, although that waiver request was denied because EPA determined that the RFS requirements alone did not "severely harm the economy of a State, a region, or the United States," a standard required by the statute. In February 2010, as part of a final rulemaking implementing the RFS as expanded by EISA, EPA waived most of the 2010 cellulosic biofuel carveout—EISA set the mandate at 100 million gallons but EPA is only requiring 6.5 million gallons, more than 90% less than scheduled by EISA. EPA cited a lack of current and expected production capacity, driven largely by a lack of investment in commercial-scale refineries.

This chapter provides a brief overview of the RFS program and discusses the process and criteria for EPA to approve a waiver petition.

INTRODUCTION

The Energy Policy Act of 2005 (EPAct, P.L. 109-58) established a renewable fuel standard (RFS), requiring the blending of biofuels (such as ethanol) in the nation's fuel supply. The Energy Independence and Security Act of 2007 (EISA, P.L. 110-140) significantly expanded this mandate. The RFS mandate has been a major impetus to the development of U.S. biofuels industries, especially the ethanol industry. As a result, ethanol production capacity and the demand for corn as a biofuel feedstock have grown dramatically over the past few years. In 2005, the United States produced 3.9 billion gallons of ethanol, requiring roughly 1.4 billion bushels of corn; in 2007, those numbers had increased to 6.5 billion gallons and 2.3 billion bushels. In 2007, roughly one-quarter of the U.S. corn crop was directed to ethanol production. In 2009, production had increased to roughly 11 billion gallons.

Increasing demand for corn for biofuels, the rise in energy prices, and other supply concerns in grain markets led to rapid increases in corn and other grain prices in 2008. These higher grain prices raised concern globally over food prices and availability. Because of these concerns, there was growing interest among some policymakers to amend or eliminate the RFS. Under the provisions of EPAct and EISA, the administrator of the Environmental Protection Agency (EPA) has the authority to waive the RFS requirements in whole or in part, in response to a petition by a state or a fuel provider, or on her own motion.

On April 25, 2008, Texas Governor Rick Perry sent a letter to EPA Administrator Stephen Johnson, petitioning for a 50% waiver from the RFS requirements.[1] In his letter, Governor Perry stated that he initiated the petition because of the negative effect of the requirements on the Texas economy and on global food prices. In August 2008 EPA denied the waiver request because the agency found that the effects of the RFS on food, feed, and fuel prices was minimal, and thus the economic effects of the RFS "could not be categorized as severe."[2]

Within the overall RFS there are secondary mandates for the use of cellulosic biofuels, biomass- based diesel fuels, and other advanced biofuels. However, questions have been raised over whether there is enough feedstock supply and production capacity to meet these carveouts, especially the cellulosic biofuel carveout. In February 2010, as part of a final rulemaking implementing the RFS as expanded by EISA, EPA waived most of the 2010 cellulosic biofuel carveout—EISA set the mandate at 100 million gallons but EPA is only requiring 6.5 million gallons, more than 90% less than scheduled by EISA. EPA cited a lack of current and expected production capacity, driven largely by delays in production plans and a lack of investment in commercial-scale refineries.[3]

Current RFS Requirements

Currently, the RFS requires the blending of 12.95 billion gallons of renewable fuel in transportation fuels in 2010, increasing to 36 billion gallons in 2022. Of this mandate, an increasing share must be met with "advanced biofuels"—biofuels produced from feedstocks other than corn starch—including cellulosic biofuel and bio-based diesel substitutes. As has been the case in previous years, in 2010 the vast majority of the mandate is expected to be met with U.S. corn ethanol (and a smaller amount of sugarcane ethanol from Brazil).

In future years, corn ethanol's share of the RFS is effectively capped at 15 billion gallons per year. The EISA amendments to the RFS specifically mandate the use of cellulosic biofuel (16 billion gallons by 2022) and biomass-based diesel fuel (1.0 billion gallons annually by 2012). However, advanced biofuels, especially cellulosic fuels, have been slow to develop and fuel production lags the EISA's mandate schedule.

Waiver Provisions

As amended by EISA, section 211 (o)(7) of the Clean Air Act[4] gives the EPA administrator the authority to waive, in whole or in part, the total volume of renewable fuel mandated by the RFS if, in her determination, there is inadequate domestic supply to meet the mandate, or if "implementation of the requirement would severely harm the economy or environment of a State, a region, or the United States."[5] Further, under certain conditions, the EPA administrator may waive (in whole or in part) the specific carve-outs for cellulosic biofuel and biomass-based diesel fuel.

General Waiver

On the petition of a state or a fuel provider, or at her own discretion,[6] the administrator may waive the overall RFS requirement for a given year. If a waiver is granted, any adjustment applies to total national requirement. Regardless of who initiates the waiver petition, all fuel suppliers' quotas would be reduced by a similar percentage. As the law is written, EPA may not waive the requirement for an individual state or supplier within a state, but must reduce the entire national mandate.

To grant the waiver, the EPA administrator must determine, in consultation with the Secretaries of Agriculture and Energy, that one of two conditions has been met:

- there is inadequate domestic renewable fuel supply; or
- implementation of the requirement would severely harm the economy or environment of a State, a region, or the United States.

However, it is unclear how EPA will interpret these criteria. In its May 1, 2007, final rule for 2007 onward,[7] EPA explicitly stated that it would not establish more specific criteria for the waiver.

> While EPA realizes that the criteria provided by the statute are quite general, the rationales of severe environmental or economic harm or inadequate domestic supply are sufficient for a basic framework upon which a petition can be built and evaluated. Each situation in which a waiver may be requested will be unique, and promulgating a list of more specific criteria in the abstract may be counter-productive.[8]

Within 90 days of receipt of the waiver petition, EPA must act to approve or disapprove the petition, after public notice and opportunity for comment.[9] If EPA does grant a waiver, the waiver expires after one year, but may be extended by the EPA administrator in consultation with the Secretaries of Agriculture and Energy.

Cellulosic Biofuel Waiver

As part of the RFS, EISA established a specific mandate for the use of cellulosic biofuels— ethanol or other fuels produced from woody or fibrous materials such as grasses, trees, etc. The cellulosic carveout was to start in 2010 at 100 million gallons, and increase to 16.0 billion gallons by 2022. Current cellulosic biofuel production is limited, with no commercial-scale plants in operation. Because of uncertainties over production capacity and cellulosic biofuel supply, in its February 2010 rulemaking, EPA reduced the cellulosic mandate from 100 million gallons to 6.5 million gallons for 2010.[10]

If the EPA administrator determines that the projected production volume of cellulosic biofuel for a given year is less than the mandated amount, she may reduce the carve-out. If she is going to do so, the administrator must reduce the required amount by November 30 of the preceding calendar year. If the administrator does reduce the mandated amount of cellulosic biofuel, she may also reduce the total volume required for that year under the RFS by an equal or lesser amount, but she is *not required* to do so. In the specific case of the 2010 level, EPA retained the overall RFS level mandated in EISA.[11]

Unlike the general wavier, only the EPA administrator may initiate a decision on a cellulosic biofuel waiver.

Biomass-Based Diesel Waiver

Similar to the cellulosic biofuel carve-out, EISA also established a specific mandate for the use of biomass-based diesel fuel. Currently, most of this fuel is "biodiesel"—a diesel fuel substitute produced from soybean oil and other vegetable oils through a process called "transesterification"—but other fuels, some of which are termed "renewable diesel," would also qualify. The biomass diesel carve-out started in 2009 at 0.5 billion gallons and increases to 1.0 billion gallons by 2012. Approximately 475 million gallons of biodiesel were produced in the United States in 2008. Because the EPA's rule was not finalized until 2010, the Agency has established a combined biomass diesel mandate for 2009 and 2010 of 1.15 billion gallons.[12]

If the EPA administrator (in consultation with the Secretaries of Energy and Agriculture) determines that there are significant market circumstances (including feedstock disruptions) "that would make the price of biomass-based diesel fuel increase significantly," the administrator may reduce the amount mandated for up to 60 days.[13] The administrator may extend the waiver for no more than an additional 60 days.

Reductions in the RFS

If the administrator waives a significant share of the above requirements, she must reduce the required volumes in all subsequent years. Specifically, she must reduce the applicable amounts in future years if she waives any of the above requirements by

- 20% or more for two consecutive years; or
- 50% or more in a single year.

For example, if the administrator reduced the overall RFS requirement by 6.0 billion gallons in both 2017 and 2018, then she would be required to reduce the total RFS requirement by 6.0 billion gallons in 2019 and beyond. The one exception is that these reductions in the RFS would not apply to the requirements before calendar year 2016.

Effects of a Waiver

Questions have been raised over how a waiver approval would affect food and fuel markets. As these markets are extremely complex, there is no simple answer. The effects of a waiver would likely depend on many factors:

- the degree to which the RFS requirements are relaxed under the waiver;
- the duration of the waiver;

- the scope of the waiver (cellulosic biofuel, biomass-based diesel, or the entire program);
- whether the waiver is extended;
- prevailing supply and prices for oil, gasoline, biofuels, and feedstock commodities; and
- whether the current tax incentive system for biofuels has been maintained or modified.[14]

In the specific case of the 2010 waiver for cellulosic biofuel, a key question is whether this waiver will undermine the credibility of the mandates in future years and undercut investment. As EPA noted in the final rule,

> In the proposal, we did a preliminary assessment of the cellulosic biofuel industry to arrive at the conclusion that it was possible to uphold the 100 million gallon standard in 2010 based on anticipated production.[15]

In the final rule, EPA does not state whether or not the agency believes there will be sufficient capacity to meet the cellulosic mandates in coming years (250 million gallons in 2011, 500 million gallons in 2012, and 1 billion gallons in 2013), although EPA states that "it is remarkable how much progress the industry has made in such a short time, and there is a tremendous growth opportunity for cellulosic biofuels over the next several years."[16] If EPA finds that mandates in later years likewise are unachievable, and if investors assume that future waivers are unavoidable, investment in cellulosic biofuel refineries may be limited.

End Notes

[1] Rick Perry, governor of Texas, *Letter to The Honorable Stephen L. Johnson, Administrator, U.S. Environmental Protection Agency*, April 25, 2008.

[2] U.S. Environmental Protection Agency, *EPA Decision on Texas Request for Waiver of Portion of Renewable Fuel Standard (RFS)*, EPA420-F-08-029, Washington, DC, August, 2008, http://www.epa.gov/otaq/renewablefuels/ 420f08029.htm.

[3] U.S. Environmental Protection Agency, *Regulation of Fuels and Fuel Additives: Changes to Renewable Fuel Standard Program*; Final Rule, EPA-HQ-OAR-2005-0161, Washington, DC, February 3, 2010, pp. 173-174, http://www.epa.gov/otaq/renewablefuels/rfs2-preamble.pdf.

[4] 42 U.S.C. 7545(o)(7).

[5] 42 U.S.C. 7545(o)(7)(A)(i).

[6] Under EPAct, only states could petition EPA for the waiver. EISA amended the RFS to allow fuel providers to file a petition, and to give the EPA administrator authority to initiate the process on her own motion.

[7] EPA, *Regulation of Fuels and Fuel Additives: Renewable Fuel Standard Program; Final Rule*, May 1, 2007. 72 Federal Register 23899-24014.

[8] 72 Federal Register 23928.

[9] As of May 1, 2008, EPA had not yet established the public comment period for the Texas waiver request.

[10] For example, EPA cited projections from the Energy Information Administration (EIA) that roughly 5 million gallons of cellulosic fuels could be produced in 2010, although some of that fuel would be cellulosic diesel fuel, which generates more credits per gallon due to its higher energy content. As finalized in the rule, the RFS requirements are based on ethanol-equivalent gallons, and the 5 million gallon number from EIA translates to roughly 6.5 million ethanol-equivalent gallons. U.S. Environmental Protection Agency, Regulation of Fuels and Fuel Additives: Changes to Renewable Fuel Standard Program; Final Rule, EPA-HQ-OAR-2005-0161, Washington, DC, February 3, 2010, p. 174, *http://www.epa.gov/otaq/*renewablefuels/rfs2-preamble.pdf.

[11] It is likely that the difference, 93.5 million gallons, will likely be met using imported sugarcane ethanol.

[12] In this way, biomass-based diesel credits generated in 2009 can be used for compliance in 2009-20 10. If the rule applied only to 2010, those credits generated by fuel blenders in 2009 would have been useless.

[13] However, the amount may not represent more than 15% of the total required amount for that year.

[14] In addition to the RFS mandate, biofuels are also subject to tax incentives for their production and use. Most significantly, fuel providers who blend ethanol into gasoline receive a tax credit of 45 cents for every gallon of ethanol they blend. In addition, small ethanol producers receive a tax credit of 10 cents per gallon on the first 15 million gallons they produce in a give year. Cellulosic biofuel producers may claim a credit of $1.01 per gallon (for cellulosic ethanol producers, this credit is reduced to 46 cents per gallon to offset the ethanol blender and small ethanol producer credits). Biodiesel and renewable diesel producers receive a tax credit of up to $1.00 per gallon of fuel they produce. For more information, see CRS Report R401 10, Biofuels Incentives: A Summary of Federal Programs, by Brent D. Yacobucci.

[15] U.S. Environmental Protection Agency, *Regulation of Fuels and Fuel Additives: Changes to Renewable Fuel Standard Program; Final Rule*, EPA-HQ-OAR-2005-0161, Washington, DC, February 3, 2010, p. 173, http://www.epa.gov/otaq/renewablefuels/rfs2-preamble.pdf.

[16] Ibid. p. 178.

In: Cars, Climate and the EPA: Issues in... ISBN: 978-1-61728-963-7
Editors: Carolynne D. Burgdoerfer

Chapter 4

ALTERNATIVE FUELS AND ADVANCED TECHNOLOGY VEHICLES: ISSUES IN CONGRESS

Brent D. Yacobucci

SUMMARY

Alternative fuels and advanced technology vehicles are seen by proponents as integral to improving urban air quality, decreasing dependence on foreign oil, and reducing emissions of greenhouse gases. However, major barriers—especially economics—currently prevent the widespread use of these fuels and technologies. Because of these barriers, and the potential benefits, there is continued congressional interest in providing incentives and other support for their development and commercialization.

Alternative fuels and advanced technology vehicles have were addressed early in the 111th Congress, as both the House and Senate versions of the American Recovery and Reinvestment Act of 2009 (H.R. 1) contained provisions supporting their development and deployment. While some of these provisions were removed in conference, the final version contains provisions for tax incentives, federal grants and loans, and other federal support for alternative fuels and advanced vehicles.

On February 3, 2010, the Environmental Protection Agency (EPA) finalized new rules for the renewable fuel standard (RFS) that was expanded by the Energy Independence and Security Act of 2007 (EISA, P.L. 110-140). In 2010, the RFS will require the use of 12.95 billion gallons of ethanol and other biofuels in transportation fuel. Within that mandate, the RFS will require the use of 0.95 billion gallons of advanced biofuels, including 6.5 million gallons of cellulosic biofuels. EISA also requires that advanced biofuels (as well as conventional biofuels from newly built refineries) meet certain lifecycle greenhouse gas reduction requirements. EPA's methodology and conclusions on various biofuels' lifecycle emissions have been controversial.

The 111th Congress is likely to further discuss alternative fuels and advanced technology vehicles as it addresses other key topics. These include their role in any federal policy to address climate change, and their role in federal energy policy. The 111th Congress may also play an oversight role in the development of major regulations: the Environmental Protection Agency's implementation of the RFS; the Department of Transportation's implementation of new fuel economy standards enacted in 2007; and the Department of Agriculture's implementation of a new Farm Bill enacted in 2008. Further, some key tax incentives for biofuels expired at the end of 2009, and may be extended in the second session of the 111th Congress.

INTRODUCTION

High levels of oil imports and high crude oil and gasoline prices in recent years have led to increased interest in the U.S. fuel supply. Recent congressional interest has focused on alternatives to petroleum, ways to improve the efficiency of the U.S. transportation sector, and ways to improve the stability and security of the petroleum supply and refining sectors.[1] From spring 2006 to summer 2008, high global oil prices (spurred by high demand) and refinery constraints in the domestic gasoline supply pushed U.S. gasoline pump prices to historic highs.

Historically, a problem in maintaining interest in alternative fuels and vehicles has been the volatility in oil and gasoline prices. Interest tends to rise as prices rise, and decline as prices dip. Arguably, statutory policies can counterbalance dips in public interest in periods of mixed market signals as seen recently. In fall 2006 and winter 2007, gasoline prices eased somewhat before rising significantly through summer 2008; and since summer 2008, petroleum and gasoline prices have fallen dramatically.

Along with fuel prices and supply, environmental concerns, especially poor air quality and concerns over the potential effects of climate change, have further raised interest in the development of alternatives to petroleum, as well as ways to use petroleum more efficiently.

Key components of federal policies to reduce petroleum consumption include the promotion of alternatives to petroleum fuels and the promotion of more efficient vehicles. This chapter provides an overview of current issues surrounding alternative fuels[2] and advanced technology vehicles[3]— issues discussed in further detail in other CRS reports referred to in each section.

MOST RECENT DEVELOPMENTS

On February 3, 2010, the Environmental Protection Agency (EPA) finalized new rules for the renewable fuel standard (RFS) that was expanded by the Energy Independence and Security Act of 2007 (EISA, P.L. 110-140). In 2010, the RFS will require the use of 12.95 billion gallons of ethanol and other biofuels in transportation fuel. Within that mandate, the RFS will require the use of 0.95 billion gallons of advanced biofuels, including 6.5 million gallons of cellulosic biofuels. EISA also requires that advanced biofuels (as well as conventional biofuels from newly built refineries) meet certain lifecycle greenhouse gas reduction requirements. EPA's methodology and conclusions on various biofuels' lifecycle emissions have been controversial.

On February 17, 2009, President Obama signed the American Recovery and Reinvestment Act of 2009 (ARRA, P.L. 111-5). Among other provisions, ARRA modifies tax credits for alternative fuel infrastructure and plug-in vehicles, expands funding for grants to states, localities, and other entities to replace older diesel engines with cleaner diesel engines or alternative fuel engines, and establishes grants for battery manufacturers and component suppliers to develop advanced vehicle batteries and system components.

On October 3, 2008, President Bush signed the Emergency Economic Stabilization Act (EESA, P.L. 110-343). Among other provisions, EESA modified and extended key tax credits for biofuels and other alternative fuels. In addition, EESA established a tax credit for the purchase of plug-in electric vehicles.[4] On June 18, 2008, President Bush signed the Food, Conservation, and Energy Act of 2008—the 2008 "Farm Bill" (P.L. 110-246). Among other provisions, the Farm Bill modified existing biofuels tax credits; most notably, the Farm Bill lowered (starting in 2009) the tax credit for ethanol produced from corn and other conventional feedstocks, and established a credit for the production of advanced biofuels produced from cellulosic matter (such as trees and perennial grasses).

Background and Analysis

Congressional Interest

Legislative Background

A combination of issues—the oil crises of the 1970s, the rise in awareness of environmental issues, concerns over energy security, increasing vehicle emissions, and high gasoline prices— have spurred interest in moving the United States away from petroleum fuels for transportation and toward alternative fuels and advanced vehicle technologies.

The Energy Policy Act of 1992

The 102nd Congress passed the Energy Policy Act of 1992 (EPAct 1992, P.L. 102-486). Among other provisions, this law requires the purchase of alternative fuel vehicles by federal agencies, state governments, and alternative fuel providers.[5] Under EPAct 1992, a certain percentage— which varies by the type of fleet (i.e., federal, state, or fuel provider)—of new passenger vehicles must be capable of operating on alternative fuels, including ethanol, methanol, natural gas, or propane. EPAct 1992 established a tax credit for the purchase of electric vehicles, as well as tax deductions for the purchase of alternative fuel and hybrid vehicles.

The Energy Policy Act of 2005

There was little congressional action on energy policy through the late 1 990s. In light of high fuel prices in the early 2000s, continued growth in domestic and global petroleum demand, and other energy policy concerns, Congress began working on comprehensive energy legislation in 2001. In the 107th Congress, an energy bill stalled in conference. The 108th Congress continued the debate over energy legislation. The conference report (H.Rept. 108-375) included provisions on vehicle tax credits, amendments to vehicle purchase requirements under the Energy Policy Act of 1992, a requirement that gasoline contain ethanol or other renewable fuels, and tax credits for ethanol and biodiesel fuels. However, this bill also stalled. Many of these topics were addressed in the 109th Congress by the Energy Policy Act of 2005 (EPAct 2005, P.L. 109-5 8), which was signed by President Bush on August 8, 2005.

The Energy Independence and Security Act of 2005

Continued pressure on energy prices and concerns over energy security after passage of EPAct 2005 led to continued discussion of energy policy in the 110th Congress. On December 19, 2007, President Bush signed the Energy Independence and Security Act of 2007 (EISA, P.L. 110-140). Among other provisions, EISA expanded the renewable fuel mandate in EPAct 2005, and significantly tightened federal fuel economy (CAFE) standards.

The 2008 Farm Bill

Biofuels—fuels produced from renewable organic matter, especially agricultural products and wastes, are seen by proponents as a key strategy for increasing energy security, promoting environmental quality, and raising farm incomes. Therefore, recent Farm Bills, especially the 2002 and 2008 Farm Bills (P.L. 107-171 and P.L. 110-246, respectively), have included titles to promote biofuels and other farm-based energy supplies. The 2002 Farm Bill established programs to promote the development of biofuels and biorefineries; the 2008 Farm Bill expanded on these programs, and expanded existing biofuels tax credits to promote the development of cellulosic fuels—fuels produced from woody or fibrous materials such as perennial grasses, fast-growing trees, and agricultural and municipal wastes.

The American Recovery and Reinvestment Act of 2009

The American Recovery and Reinvestment Act of 2009 (P.L. 111-5) includes several key provisions supporting alternative fuels and advanced technology vehicles. These include tax credits for the purchase of small electric vehicles, grants to states, localities, and other entities to replace older diesel engines with new, clean diesel or alternative fuel engines, and grants to battery manufacturers and part suppliers to develop batteries and system components for advanced vehicles (e.g., hybrids, plug-in electric vehicles). A provision contained in the Senate version to expand U.S. Department of Agriculture (USDA) biorefinery grants was dropped by the Conference Committee. For a more detailed comparison of the House, Senate, and Conference versions of the bill, see Table 1.

Other Legislation

Other laws affecting alternative fuel and advanced technology vehicles include the Energy Policy and Conservation Act (P.L. 94-163), which established fuel economy standards for passenger cars and light trucks;[6] the 1990 Amendments to the Clean Air Act (P.L. 10 1-549), which require cities with significant air quality problems to promote low emission vehicles; highway authorization bills, including P.L. 109-59 and P.L. 105-178, which established and reaffirmed tax incentives for ethanol and other fuels; and numerous laws that authorize federal research and development on alternative fuels, advanced technologies, and enabling infrastructure, such as alternative fuel pumps. The Emergency Economic Stabilization Act of 2008 (P.L. 110-343) modified and extended key tax incentives for biodiesel and other alterative fuels.

Current Issues

Recent events have renewed interest in alternative fuels and advanced vehicles. For example, high pump prices for gasoline and diesel fuel through summer 2008 raised concerns over fuel conservation and energy security, including U.S. dependency on oil imports. In light of this, there is growing interest in more efficient vehicles or vehicles that abandon the use of petroleum altogether. This is especially true as the rapid growth in the sales of light trucks—these include sport utility vehicles (SUVs), mini-vans, and pickups, which tend to have lower fuel economy than passenger cars—through the mid-2000s lowered the overall fuel economy of the new vehicle fleet. EISA requires an increase in fuel economy from passenger cars and light trucks to 35 miles per gallon (mpg) combined in 2020 from roughly 24 mpg today.

Ongoing technological developments in hybrid vehicles, ethanol fuel, fuel cells, and hydrogen fuel have raised key policy questions. These questions include whether more generous tax incentives for hybrid and/or fuel cell vehicles should be established, the costs and environmental impacts associated with production of ethanol or hydrogen as major transportation fuels, and whether research and development funds should be focused on such potentially high-risk technologies as fuel cells or on near-term technologies, such as hybrids.

Gasoline prices have spiked and the gasoline supply system has faced disruptions several times in recent years, driven by various factors, including

- hurricanes along the Gulf Coast in the fall of 2005;

- high crude prices, issues with refining capacity, and concerns about ethanol supply in spring 2006;
- historic high crude oil and gasoline prices returned to historic highs in 2007 and 2008.

These price surges and supply disruptions raised congressional interest in alternatives to petroleum. Coupled with concerns over the environmental impact of petroleum and other fossil fuels, congressional interest in alternatives remains strong, even though oil and gasoline prices declined during the second half of 2008.

Fuel Tax Incentives

There are three key tax incentives for alternative fuels: (1) a tax credit for conventional ethanol of $0.45 per gallon,[7] (2) a tax credit for biodiesel and renewable diesel of $1.00 per gallon,[8] and (3) a credit of $0.50 per gallon for the retail sale of alternative fuels other than ethanol and biodiesel (e.g., LPG). In addition, there are tax credits for small ethanol and biodiesel producers ($0.10 per gallon), and a tax credit for the production of cellulosic biofuels (up to $1.01 per gallon, depending on the fuel).[9] The credits for the retail sale of alternative fuels and for biodiesel and renewable diesel expired at the end of 2009. As of January 2010, these credits had not been extended, although there is congressional interest in doing so.

In general, there is ongoing interest in tax incentives for the production and purchase of alternative fuels. Supporters of this approach argue that the market favors conventional fuels, and that the widespread infrastructure and nearly ubiquitous use of conventional fuels in automobiles makes it difficult for alternative fuels to compete without economic incentives. The American Jobs Creation Act of 2004 (P.L. 108-357) replaced a previous excise tax exemption for ethanol- blended fuels with a tax credit. The credit was valued at $0.51 per gallon in 2008 and is expected to be reduced to $0.45 per gallon for 2009.

In addition to the credit for ethanol-blended gasoline, there has been interest in promoting biodiesel fuel. P.L. 108-357, and subsequent amendments, provides a tax credit of $1.00 per gallon for the sale and use of biodiesel. P.L. 109-58 expanded the credit to include "renewable diesel," which is produced from a different process than biodiesel and results in a fuel with somewhat different chemical characteristics. In guidance on the tax credit, the Internal Revenue Service ruled that renewable diesel includes synthetic diesel fuel produced from vegetable oils at petroleum refineries. [10] Most biodiesel producers are small plants, and many biodiesel producers were concerned that this decision could lead to a shift away from biodiesel production to renewable diesel production at large refineries, although this effect seems limited. The Emergency Economic Stabilization Act of 2008 (EESA) included the following language amending the renewable diesel tax credit: "such term does not include any fuel derived from coprocessing biomass with a feedstock which is not biomass."[11] Presumably, this provision is intended to limit the production of renewable diesel eligible for the tax credit at petroleum refineries. EESA also extended the biodiesel and renewable diesel credits through the end of 2009. These credits have not been reinstated in 2010, although there is congressional interest in doing so.

Ethanol and MTBE

Outside of tax incentives, ethanol has been of key interest in recent Congresses, especially in its role as an alternative to gasoline and to MTBE (methyl tertiary butyl ether). [12] MTBE and ethanol were used (among other purposes) to meet previous Clean Air Act requirements that reformulated gasoline (RFG), sold in the nation's worst ozone nonattainment areas, contain oxygen to improve combustion. Under the RFG program, areas with "severe" or "extreme" ozone pollution (90 counties with a combined population of 64.8 million in 2009[13]) must use reformulated gas; areas with less severe ozone pollution may opt into the program as well, and many have. In all, portions of 17 states and the District of Columbia use RFG, and about 30% of the gasoline sold in the United States is RFG, according to the Environmental Protection Agency (EPA).[14]

Before amendment by the Energy Policy Act of 2005, the Clean Air Act required that RFG contain at least 2% oxygen by weight.[15] Refiners met this requirement by adding a number of ethers or alcohols, any of which contains oxygen and other elements. Until about 2003, the most commonly used oxygenate was MTBE because it was cheaper and easier to use than competing oxygenates. In 1999, 87% of RFG contained MTBE, a number reduced to about 46% in 2004, according to EPA. MTBE has also been used since the late 1970s in non-reformulated gasoline as an octane enhancer, at lower concentrations. As a result, gasoline with MTBE has been used throughout the United States, whether or not an area has been subject to RFG requirements.

MTBE contamination creates taste and odor problems in water at very low concentrations, and some animal studies indicate MTBE may pose a cancer risk to humans. MTBE leaks, generally from underground gasoline storage tanks, have been implicated in numerous incidents of ground water contamination. For these reasons, 25 states have taken steps to ban or limit its use, according to the Renewable Fuels Association.[16] The most significant of the bans (in California and New York) took effect at the end of 2003, leading many to suggest that Congress revisit the issue to modify the oxygenate requirement and set more uniform national requirements regarding MTBE and its potential replacements, principally ethanol.

Support for eliminating the oxygenate requirement on a nationwide basis was widespread among states, the petroleum industry, and some environmental groups. In general, these stakeholders concluded that gasoline can meet the same low-emission performance standards as RFG without the use of oxygenates. But agricultural interests presented a potential obstacle to enacting legislation to remove the oxygen requirement. According to USDA, roughly 20% of the nation's corn crop was used in 2006/2007 to produce the competing oxygenate, ethanol.[17] If MTBE use were reduced or phased out, but the oxygen requirement remained in effect, ethanol use would have soared, increasing demand for corn. Conversely, if the oxygen requirement were repealed, not only would MTBE use decline, but so, likely, would demand for ethanol. Thus, some Members of Congress and governors from corn-growing states took a keen interest in MTBE legislation and related oxygenate requirements.

To help promote the market for ethanol if the oxygen standard were eliminated, a renewable fuel standard (RFS) was suggested. This would require that all gasoline contain ethanol or other renewable fuel. This concept was supported by agricultural interests, the oil industry, and some environmental groups. Opponents included states that do not produce ethanol, due to fears that the mandate could raise gasoline prices.

The Energy Policy Act of 2005 (P.L. 109-58) contains numerous MTBE and ethanol provisions. It repealed the Clean Air Act requirement to use MTBE or other oxygenates. In place of this requirement, the law established a renewable fuel standard. Under the RFS, annual gasoline supply was required to contain 7.5 billion gallons of ethanol or other renewable fuel by 2012. To prevent "backsliding" on air quality, the law requires that reductions in emissions of toxic substances achieved by RFG be maintained, and it authorizes funds for MTBE cleanup. [18] The Energy Independence and Security Act of 2007 (P.L. 110-140) expanded this mandate, requiring the use of 9.0

billion gallons of renewable fuels in transportation fuel[19] in 2008, and 36 billion gallons in 2022.

The Renewable Fuel Standard (RFS)

On February 3, 2010, the Environmental Protection Agency (EPA) finalized rules for the RFS as expanded by the Energy Independence and Security Act of 2007 (EISA, P.L. 110-140).[20] As mandated by EISA, the rule will require the use of 12.95 billion gallons of renewable fuels in transportation fuels in 2010. Most of this mandate will be met using ethanol produced from corn, although within the larger RFS mandate, there are carve-outs for cellulosic biofuels, biomass- based diesel substitutes, and other advanced biofuels.

One area of controversy is EPA's conclusion about the greenhouse gas impacts of biofuels. As part of its expansion of the RFS, EISA requires that all advanced biofuels, as well as conventional biofuels from new refineries, have reduced greenhouse gas emissions relative to gasoline. In its proposed rule, EPA found that many fuel pathways did not meet the threshold requirements in EISA. However, its methodology was criticized by biofuels supporters. In the final rule, EPA modified its methodology to reflect some of those comments. However, some biofuels opponents counter that the final rules went too far in the opposite direction.[21]

A key component of the expanded RFS is a requirement starting in 2010 that a growing portion of the RFS be met using cellulosic biofuels (see "Cellulosic Biofuels"). Under EISA, the cellulosic biofuel mandate begins at 100 million gallons in 2010 and increases to 16 billion gallons by 2022. However, EPA concluded that U.S. production capacity will be well below 100 million gallons in 2010: Using its authority under EISA to waive parts of the RFS, EPA set the cellulosic biofuel mandate at 6.5 million gallons (ethanol equivalent) for 2010.[22]

> While we proposed that the cellulosic biofuel standard would be set at the EISA specified level of 100 million gallons for 2010, based on analysis of information available at this time, we no longer believe the full volume can be met.... we have found that many of the projects that served as the basis for the proposal have been put on hold, delayed, or scaled back. At the same time, there have been a number of additional projects that have developed and are moving forward.... the timing for many of the projects indicates that

> while few will be able to provide commercial volumes for 2010, an increasing number will come on line in 2011, 2012, and 2013.... 5 million gallons (6.5 million ethanol equivalent) represents a reasonable, yet achievable level for the cellulosic standard for 2010.[23]

By November 2010 EPA will need to publish the required fuel levels for 2011, including the cellulosic biofuel mandate. Of key interest will be whether EPA believes that the level set in EISA for 2011—250 million gallons—will be met, or whether EPA will again need to use its waiver authority.

Cellulosic Biofuels[24]

Ethanol, the most significant biofuel in the United States, is usually produced from corn. However, corn is a key animal feed, and is also used for human consumption. Further, corn is a resource-intensive crop, requiring significant use of chemical fertilizers and generally grown on prime farmland. There is growing interest in developing biofuels that require less energy to produce and have a smaller environmental footprint.

Biofuels produced from cellulosic materials such as fast-growing trees, prairie grasses, or agricultural wastes are seen as a potential strategy for reducing the environmental impact of biofuels while expanding the United States' ability to displace petroleum fuels. The potential supply of these feedstocks is abundant, which is why it is expected that future expansion of the U.S. biofuels industry will be in this area.

However, breaking down cellulose and converting it into fuel requires complex chemical processing. Starches (such as corn) and sugars (such as cane sugar) are easily fermented into alcohol, while cellulose must be broken down into sugars or starches through enzymatic or thermochemical processes before fermentation. Alternatively, biomass can be converted into synthesis gas,[25] which can then be used to produce fuels. Regardless of the pathway, processing cellulose into fuels is currently prohibitively expensive relative to other conventional and alternative fuel options. Therefore, R&D has focused on lowering the costs of enzymatic and other processing techniques.

Further, questions remain about the feasibility of these fuels, as well as the ultimate environmental footprint—many of the proposed feedstocks have never been grown on a large scale. Therefore, R&D is also focused on increasing the yield of potential biofuel crops, developing harvesting techniques, and finding ways to limit the environmental impact of dedicated energy crops.

The Energy Policy Act of 2005 included provisions to promote the development of cellulosic biofuels. These include an authorization for increased research and development funding at the Department of Energy; grants, loans, and loan guarantees for the development of cellulosic biofuels; per-gallon incentives for the first 1 billion gallons of domestic production;[26] and a mandate that gasoline contain at least 250 million gallons of cellulosic ethanol annually starting in 2013.

On December 20, 2006, President Bush signed the Tax Relief and Health Care Act of 2006 (P.L. 109-432). Among other provisions, this tax law established a 50% depreciation allowance for cellulosic ethanol plants placed in service before January 1, 2013, subject to certain limitations.

The Energy Independence and Security Act of 2007 expanded the renewable fuel mandate in EPAct 2005, and established specific requirements for "advanced biofuels"—defined as fuels produced from feedstocks other than corn starch, and with 50% lower lifecycle greenhouse gas emissions than petroleum fuels. (See "The Renewable Fuel Standard (RFS).") Of the 36 billion gallons of renewable fuel required in 2022, 21 billion gallons must be advanced biofuels; within that mandate, there are specific carve-outs for cellulosic biofuels and biomass-based diesel fuels.

Ethanol Imports

Corn growers and ethanol producers are supportive of the renewable fuel standard because of its implications for higher corn and ethanol prices. However, concern over ethanol imports has grown among some stakeholders. Because of lower production costs and the availability of government incentives, ethanol prices in Brazil and some other countries can be significantly lower than in the United States. To offset the U.S. tax incentive that all ethanol (imported or domestic) receives, most imports are subject to a relatively small 2.5% ad valorem tariff, but more significantly an added duty of $0.54 per gallon. This added duty effectively negates the tax incentive for covered imports and has been a significant barrier to fuel ethanol imports.

However, under certain conditions imports of ethanol from Caribbean Basin Initiative (CBI) countries are granted duty-free status.[27] This is true even if the ethanol was produced in a non- CBI country. In this scenario, the ethanol is produced in another country (historically Brazil or a European country), dehydrated in a CBI country, then shipped to the United States. In recent years, these imports have reached as high as 5% of the U.S. ethanol market. This avenue for imported ethanol to avoid the tariff has been criticized by some stakeholders, including some Members of Congress. With the establishment of a renewable fuel standard, as well as high U.S. gasoline and ethanol prices, there may be more interest in importing ethanol, either through CBI countries or directly from ethanol producers.

In addition to the concerns over imports of duty-free ethanol from CBI countries, there is growing concern that a large portion of ethanol otherwise subject to the duties is being imported duty-free through a "manufacturing drawback."[28] If a manufacturer imports an intermediate product, then exports the finished product or a similar product, then that manufacturer may be eligible for a refund (drawback) of up to 99% of the duties paid. There are special provisions for the production of petroleum derivatives.[29] In the case of fuel ethanol, the imported ethanol is used as a blending component in gasoline, and jet fuel (considered a like commodity) is exported to qualify for the drawback. Some critics estimate that as much as 75% or more of the duties were eligible for the drawback in 2006.[30] Therefore, critics question the effectiveness of the ethanol duties and the CBI exemption.

On December 20, 2006, President Bush signed the Tax Relief and Health Care Act of 2006 (P.L. 109-432). Among other provisions, the act extended the duty on imported ethanol through December 31, 2008, but did not address the duty drawback provisions or the CBI preference. The 2008 Farm Bill further extended the duty through December 31, 2010.

Vehicle Purchase Requirements

The Energy Policy Act of 1992 established mandatory alternative fuel vehicle purchase requirements for various vehicle fleets.[31] Under the law, 75% of the passenger vehicles purchased by federal and state vehicle fleets must be capable of operating on alternative fuels; 90% of the vehicles purchased by alternative fuel providers must be alternative fuel vehicles.[32]

The alternative fuel vehicle provisions of EPAct 1992 have been criticized as ineffective because, while EPAct 1992 requires the purchase of vehicles capable of operating on alternative fuels, it did not mandate the use of alternative fuels. In most cases, the vehicles purchased to meet the requirement are dual-fuel vehicles (i.e., they can operate on either a conventional fuel or an alternative fuel). Those vehicles are primarily fueled using gasoline, because gasoline tends to be less expensive and more widely available than alternative fuels because the infrastructure to provide alternative fuels is limited compared with the existing infrastructure for gasoline and diesel fuel.

In addition, despite the vehicle purchase mandate, in previous years many agencies failed to meet their statutory obligation. As a result, in 2002 the Center for Biological Diversity filed a lawsuit with the U.S. District Court for the Northern District of California. In July 2002, the court ruled that several federal agencies failed to meet their quotas and ordered those agencies to prepare reports on their compliance with EPAct, which those agencies have completed.[33] Since that time, most agencies have complied with the requirement; in FY2007, the most recent year data are available, all covered federal fleets met the requirement.[34]

The Energy Policy Act of 2005 (Section 701) modified the requirements for EPAct 1992 compliance. All dual-fuel vehicles purchased to meet the EPAct quotas are required to operate on alternative fuels, unless an agency is granted a waiver by the Secretary of Energy. However, it is unclear whether this requirement will significantly affect federal agency alternative fuel use. The Secretary of Energy is required under the law to conduct a study of the effectiveness of the EPAct requirements. Further, Section 703 of EPAct 2005 allows state and fuel provider fleets to petition the Department of Energy (DOE) to waive the vehicle purchase requirement if the fleet certifies other fuel-saving measures (e.g., using higher-efficiency conventional vehicles or hybrids).

On January 28, 2008, President Bush signed the National Defense Authorization Act for Fiscal Year 2008 (P.L. 110-181). Among other provisions, the law amends the definition of "alternative fuel vehicle." Under the new definition, fleets covered by EPAct 1992 will be granted credits for the purchase of hybrid, advanced diesel,[35] and fuel cell vehicles, in addition to those alternative fuel vehicles already allowed.

In addition to the requirements for federal, state, and fuel provider fleets, EPAct 1992 grants the DOE the authority to extend the requirements to local government and private fleets. However, as of 2002, DOE had not made a determination on requirements for local and private fleets. As part of the above lawsuit, the Center for Biological Diversity also asked the court to force DOE to promulgate new rules. In ruling on the above case, the U.S. District Court for the Northern District of California ordered DOE to establish a timeline for a new rulemaking. DOE compiled a timeline and, on March 4, 2003, it issued a rulemaking determining that such a program would not promote the goals of

EPAct, neither reducing dependence on foreign oil nor leading to greater use of alternative fuel vehicles.[36]

The American Recovery and Reinvestment Act of 2009 (H.R. 1) appropriated $300 million to the General Services Administration for the purchase of vehicles with high fuel economy, including hybrid, plug-in hybrid, and pure electric vehicles.

Vehicle Purchase Tax Incentives

Some supporters of alternative fuel and advanced technology vehicles argue that tax incentives for the purchase of vehicles and fuels are more effective than any purchase mandate. In addition to the mandatory purchase requirements, EPAct 1992 established tax incentives for the purchase of electric vehicles and "clean-fuel vehicles," including alternative fuel and hybrid vehicles. The Energy Policy Act of 2005 (Section 1341) significantly expanded and extended the vehicle purchase incentives, establishing tax credits for the purchase of fuel cell, hybrid, alternative fuel, and advanced diesel vehicles. For passenger vehicles, the credit is worth as much as $3,400 for hybrids and advanced diesels, and as much as $4,000 for alternative fuel vehicles, depending on vehicle attributes. The expiration date for the incentives also varies depending on the technology.[37]

In the case of hybrid and advanced diesel vehicles, the number of vehicles eligible for the credit is limited for each vehicle manufacturer. Starting the second calendar quarter after a manufacturer sells the 60,000th vehicle eligible for the credit, the credit for that manufacturer's vehicles is reduced. Currently, only Toyota and Honda have sold enough vehicles to trigger a phaseout. For Toyota (and Lexus) hybrids the credit phased out after September 30, 2007. Honda's phaseout began January 1, 2008, and the credit for Honda vehicles reached zero on January 1, 2009. Other manufacturers have yet to hit the 60,000 vehicle mark, although in general the credit expires at the end of 2010.[38]

The Emergency Economic Stabilization Act of 2008 established a tax credit for the purchase of plug-in vehicles, both pure electric vehicles and plug-in hybrids (i.e., gasoline/electric hybrid vehicles that can fuel on gasoline or be recharged from the electric grid.) For passenger vehicles, the credit is a maximum of $7,500, depending on the vehicle's battery capacity. After sales of vehicles eligible for the credit exceed a total of 250,000 for all manufacturers, the credit is phased out.

The American Recovery and Reinvestment Act of 2009 (P.L. 111-5) significantly modified the plug-in credits: the law eliminated the credit for vehicles above 14,000 pounds after 2009; established a credit of up to $2,500 for 2-wheeled, 3-wheeled, and low-speed 4-wheeled plug-in vehicles; and established a credit of up to $4,000 for the conversion of existing vehicles to run on battery power. The law also allows purchasers to claim the plug-in, alternative fuel vehicle, and hybrid tax credits even if they are subject to the Alternative Minimum Tax (AMT)—previously, taxpayers subject to the AMT could not claim these credits.

Biodiesel

Biodiesel is a synthetic diesel fuel produced from oils, including soybean and canola oils, animal fats, and recycled cooking grease. It can be blended with conventional diesel fuel and used in diesel engines with few or no modifications. Further, with some engine modifications, it can be used in a nearly pure form. Because biodiesel can displace conventional diesel without the use of new (and in many cases costly) vehicles, there is growing interest in its use. Further, because it can be produced from agricultural products, farmers (especially soybean and canola farmers) and some environmentalists have a keen interest in its development as a way to promote rural economies, reduce agricultural wastes, and limit greenhouse gas emissions. However, biodiesel production is currently expensive: wholesale biodiesel from virgin oils can cost up to two times more than conventional No. 2 diesel; biodiesel from recycled grease is less expensive but still costs considerably more than conventional diesel.

The cost barriers for biodiesel production have generated interest in providing tax incentives for biodiesel, in the form of either a production tax credit or an excise tax exemption, or both. Further there is interest in developing new technologies to help reduce production costs. However, the organic oils used as raw materials are one of the largest costs in production. Therefore, to significantly reduce biodiesel production costs, the costs of soybean oil and other oils would need to decrease substantially.

As was stated above, the American Jobs Creation Act, as amended, provided a tax credit of up to $1.00 per gallon for the sale and use of biodiesel. In addition, the law provided an excise tax credit for biodiesel blends (i.e., biodiesel and conventional diesel). Producers are eligible for one credit or the other, but not both (see "Fuel Tax Incentives" above). These credits were set to expire at the end of 2006; the Energy Policy Act of 2005 (P.L. 109-58) extended these credits through 2008. Further, EPAct 2005 established a credit of $0.10 per gallon for small agribiodiesel producers, and a $1.00-per-gallon credit for "renewable diesel"—diesel fuel produced from biomass through a different process than the biodiesel production process. The Emergency Economic Stabilization Act (P.L. 110-343) further extended these credits through the end of 2009. As of January 2010, they had not been extended. While these tax credits generally do not make biomass-based diesel fuels less expensive than conventional diesel, they do help make them more cost-competitive.

Hydrogen and Fuel Cells

Over the past few years, interest has grown substantially in hydrogen fuel and fuel cells. Hydrogen fuel can be produced using any energy source, and has thus been touted as a way to limit dependence on energy imports. Further, when hydrogen is used in a fuel cell (a device that produces electricity by converting hydrogen to water), mostly heat and water are produced, drastically reducing or eliminating vehicle emissions. However, hydrogen fuel production is currently very expensive, as are fuel cells. In addition, depending on the original fuel source, overall fuel-cycle emissions can be a key concern.[39]

Because of the potential benefits from hydrogen and fuel cells, and because of the existing technical and cost barriers to their commercialization, the Bush Administration strongly supported research and development (R&D). In January 2002, the Bush Administration announced the FreedomCAR initiative, which promotes cooperative R&D between the "Big Three" American auto manufacturers (Chrysler, Ford, and General Motors) and the federal government. While the partnership is conducting research on many automotive technologies, hydrogen and fuel cell vehicles have been a key focus. Further, in his January 2003 State of the Union address, President Bush announced the Hydrogen Fuel Initiative, which increased federal spending on hydrogen fuel and stationary fuel cell R&D. Overall, the President requested $1.8 billion between FY2004 and FY2008 for both initiatives, including a $720 million increase in funding from earlier appropriations. Over that time, Congress appropriated a total of $1.4 billion for the initiatives.[40] The Energy Policy Act of 2005 authorized a total of $3.3 billion through FY2010 for fuel cell and hydrogen R&D.[41]

Opponents of the initiatives argue that hydrogen fuel and fuel cells may never be commercialized and that the initiatives draw funding away from near-term technologies such as hybrid vehicles. Further, some argue that research and development alone will not reduce petroleum dependence and that Congress should instead consider tightening fuel economy standards for all vehicles. As noted earlier, Congress did tighten fuel economy standards for all vehicles in the Energy Independence and Security Act of 2007 (P.L. 110-140).

Hybrid Vehicles

Hybrid gasoline/electric (and diesel/electric) vehicles are becoming increasingly popular in the United States. Hybrids combine a gasoline (or diesel) engine with an electrical motor system to improve efficiency. If their use becomes more widespread, they could help improve the overall efficiency of the vehicle fleet and could help limit oil consumption. Further, they could do so without significant changes to existing infrastructure, which has been a key barrier to the expanded use of alternative fuel vehicles. By January 2010, BMW, Ford, General Motors, Honda, Nissan, Mazda, Mercedes-Benz, and Toyota offered vehicles with hybrid powertrains. At the present time, only hybrid passenger cars, SUVs, and pickups are available in the United States, but hybrid versions of other vehicle models and classes are expected in the near future.

Because of their energy and environmental benefits, some states have provided drivers of hybrid vehicles an exemption from high occupancy vehicle (HOV) lane requirements. Under the 1998 surface transportation bill (which expired on September 30, 2003), states had the authority to grant HOV exemptions for so-called "Inherently Low Emission Vehicles" (ILEVs). The ILEV standard requires that a vehicle have no evaporative emissions, a standard that is not met by any current hybrid. However, because of the reduced emissions and improved fuel economy of hybrid vehicles, there has been congressional interest in explicitly granting states the right to exempt them from HOV lane requirements. While not addressing hybrids directly, the final version of the 2005 surface transportation reauthorization act (P.L. 109-59) permits states to exempt certain high-efficiency vehicles from HOV restrictions.

Further, as was stated above, the Energy Policy Act of 2005 and the Emergency Economic Stabilization Act of 2008 expanded the incentives for the purchase of hybrid and plug-in hybrid vehicles (see "Vehicle Purchase Tax Incentives" above).

Alternative Fuel and Advanced Vehicle Technology Provisions in the American Recovery and Reinvestment Act of 2009

Table 1. Comparison of Alternative Fuel and Advanced Vehicle Technology Provisions in H.R. 1

Topic	House Version	Senate Version	Conference Version	Comments
Advanced Battery Manufacturing	Appropriates $1 billion for loan guarantees for advanced battery manufacturing, and $1 billion for grants for advanced battery manufacturing facilityies under Secs. 135 and 136 of EISA, respectively	Appropriates $2 billion for grants for advanced battery manufacturers	Similar to Senate provision	The House version makes direct reference to authorizations under EISA, while the Senate and Conference versions do not
Grants to States for Transportation Electrification	Appropriates $200 million for transportation electrification projects under Sec. 131 of EISA	Appropriates $200 million for transportation electrificationa	Appropriates $400 million for transportation electrificationb	Sec. 131 of EISA authorizes electrification grants for a variety of transportation modes, including highway vehicles, airport ground support vehicles, and ships
Grants for Advanced Vehicles	Appropriates $400 million for grants through the Clean Cities program for the purchase of alternative fuel and advanced technology vehicles under Sec. 721 of EPAct 2005	Appropriates $350 million through the Clean Cities program for the purchase of alternative fuel and fuel cell vehicles[a]	Similar to House provision, but appropriates $300 million for Sec. 721 of EPAct 2005[b]	Sec. 721 of EP Act 2005 authorizes grants to states, localities, and metropolitan transit agencies for the purchase of alternative fuel and advanced technology vehicles

Table 1. (Continued)

Topic	House Version	Senate Version	Conference Version	Comments
GSA Purchases of Fuel Efficient and Alternative Fuel Vehicles	Appropriates $600 million for the purchase of alternative fuel and fuel efficient vehicles for the federal fleet	Appropriates $300 million for the purchase of vehicles with higher fuel economy (including advanced technology vehicles)	Similar to Senate provision	
Diesel Emissions Reduction	Appropriates $300 million for diesel emission reduction grants under Title VII, Subtitle G of EPAct 2005	Similar to House provision	Similar to House provision	EPA funding for this program in recent years has been around $50 million annually
Expansion of Alternative Fuel Refueling Infrastructure Tax Credit	For 2009 and 2010, expands the Current credit for the installation of alternative fuel refueling infrastructure: increases percentage credit for retail installations to 50% and maximum credit to $50,000; for hydrogen retail infrastructure, maintains 30% credit but increases maximum to $200,000; increases residential credit to $2,000	Similar to House provision, but expands credit to include electric vehicle recharging stations, if certain conditions are met	Similar to House provision	Sec. 1342 of the EPAct 2005 established a tax credit for the installation of infrastructure to deliver alternative fuels —defined as ethanol, natural gas, liquefied petroleum gas, hydrogen, or fuels containing at least 20% biodiesel; the credit is 30% of the installation cost, up to $30,000 for retail installations, or up to $1,000 for residential installations
Modification of Tax Credit for the Purchase of Plug-in Vehicles	No provision	Modifies the existing tax credit for plug-in vehicles to allow a 10% credit, up to $4,000 for the purchase of 4-wheeled low-speed electric vehicles, as well as 2- and 3-wheeled	Modifies the existing tax credit to cap the per-vehicle credit at $7,500 for light-duty vehicles and heavy-duty vehicles up to 14,000 pounds gross weight; replaces the 250,000	Section 205 of EESA established a tax credit for the purchase of a new plug-in vehicle; the credit is based on the battery capacity of the vehicle, and is capped at $7,500 for light-duty

Table 1. (Continued)

Topic	House Version	Senate Version	Conference Version	Comments
		electric vehicles; establishes a 10% credit, up to $4,000 for the conversion of an existing vehicle to battery power	total vehicle limit for phase-out of the credit with a 200,000 per-manufacturer limit; eliminates the credit for heavier vehicles; establishes a credit of up to $2,500 for low-speed 4-wheel vehicles, as well as 2- and 3-wheeled electric vehicles; establishes a credit of up to $4,000 for the conversion of an existing vehicle to battery power; allows plug-in credit (as well as other alternative fuel and advanced vehicle credits) as a personal credit against the Alternative Minimum Tax (AMT)	vehicles and up to $15,000 for the heaviest vehicles; when total U.S. sales of vehicles eligible for the credit reaches 250,000, the credit begins to phase out; currently purchasers may not claim the plug-in credit and related credits for alternative fuel and advanced technology vehicles if they are subject to the AMT
Biorefinery Grants and Loan Guarantees	No Provision	Appropriates $200 million for 2008 Farm Bill Biorefinery Assistance Program	No provision	As yet unfunded, this program would provide financial assistance for the development, construction, and conversion of plants to produce advanced biofuels

Source: CRS Analysis of: H.R. 1 as passed by the House and Senate; Senate Committee on Appropriations, *Appropriations Provisions of the American Recovery and Reinvestment Act,* February 6, 2009; Joint Explanatory Statement of the Committee of Conference on H.R. 1; H.Rept. 111-16.

a. Senate report language specifies this appropriation.

b. Conference report language (Joint Explanatory Statement) specifies this appropriation.

For Additional Reading

California Energy Commission. (1999). *ABCs of AFVs: A Guide to Alternative Fuel Vehicles.* Sacramento, CA. November.

Electric Drive Transportation Association. (2010). *Technology/Vehicle Overview.* Washington, DC. Updated January. *http://www.elect* ricdrive.org

Fuel Cells (2000). *Online Fuel Cell Information Center*. Washington, DC. Updated January 2010. http://www.fuelcells.org/

Methanol Institute. (2009). *Methanol Institute Homepage*. Washington, DC. Updated November. http://www.methanol.org/

National Biodiesel Board. (2010). *Biodiesel Basics.* Jefferson City, MO. Updated January. http://www.biodiesel

National Hydrogen Association. (2010). *General Information.* Washington, DC. Updated January. *http://www.hydrogenassociation.org* /general/index.asp

Natural Gas Vehicles for America. (2009). *About NGVs.* Washington, DC. Updated July. http://www.ngvc.org/about_ngv/index.html

Propane Education and Research Council. (2009). *Propane Vehicle Fleets.* Washington, DC. Updated December. *http://www.propanecouncil* .org/trade

Renewable Fuels Association. *(2009). Ethanol Industry Outlook*. Washington, DC. February 2009. http://www.ethanolrfa.org/industry

U.S. Department of Energy. (2009). Office of Energy Efficiency and Renewable Energy. *Alternative Fuels Data Center.* Washington, DC. Updated December. http://www.eere.energy

U.S. Department of Energy. (2002). *Fuel Cell Report to Congress.* Washington, DC. February 2003. ——. *National Hydrogen Energy Roadmap*. Washington, DC. November.

U.S. Government Accountability Office (formerly Government Accounting Office). (1992). Energy Policy Act of: Limited Progress in Acquiring Alternative Fuel Vehicles and Reaching Fuel Goals. Washington, DC. February 2000. RCED-00-59.

——. (2000). *Tax Incentives for Petroleum and Ethanol Fuels.* Washington, DC. September. RCED-00-30 1R.

U.S. Internal Revenue Service. (2008). *Hybrid Cars and Alternative Fuel Vehicles.* Updated October 3,. *http://www.irs.gov/newsroom/article* /0,,id=157632,00.html

End Notes

[1] For more information on petroleum supply and prices, see CRS Report RL32530, *World Oil Demand and its Effect on Oil Prices*, by Robert Pirog. For more information on legislative proposals to help mitigate high gasoline prices, see CRS Report RL33521, *Gasoline Prices: Causes of Volatility and Congressional Response*, by Carl E. Behrens and Carol Glover.

[2] Alternative fuels are fuels produced from sources other than petroleum, including natural gas, coal-derived fuels, agriculture-based ethanol and biodiesel, and hydrogen.

[3] Advanced technology vehicles are vehicles that use technologies other than (or in addition to) an internal combustion engine, including electric vehicles, fuel cell vehicles, and hybrids.

[4] Tax credits were established in 2005 (by the Energy Policy Act of 2005—P.L. 109-5 8) for the purchase of hybrid, lean-burn, alternative fuel, and fuel cell vehicles.

[5] Alternative fuel providers are businesses that sell or distribute alternative fuels.

[6] For more information on fuel economy standards, see CRS Report R40 166, *Automobile and Light Truck Fuel Economy: The CAFE Standards*, by Brent D. Yacobucci and Robert Bamberger.

[7] Through 2008, the tax credit was valued at $0.51 per gallon. The 2008 Farm Bill lowered the credit to $0.45 per gallon in the first year after U.S. ethanol supply exceeds 7.5 billion gallons. Through October 2008, annual U.S. ethanol consumption had already exceeded 7.8 billion gallons: Renewable Fuels Association, *The Industry - Statistics*, 2008 Monthly U.S. Fuel Ethanol Production/Demand, Washington, DC, http://www.ethanolrfa.org/industry

[8] Through 2008, the credit for biodiesel produced from recycled materials was $0.50 per gallon. EESA eliminated the distinction between biodiesel fuels produced from different feedstocks.

[9] For more information on tax and non-tax incentives for ethanol and biodiesel, see CRS Report R401 10, *Biofuels Incentives: A Summary of Federal Programs*, by Brent D. Yacobucci.

[10] U.S . Internal Revenue Service, Notice 2007-37: Renewable Diesel, April 23, 2007.

[11] P.L. P.L. 110-343, Division B, Sec. 202.

[12] For additional background on the MTBE issue, see CRS Report RL32787, *MTBE in Gasoline: Clean Air and Drinking Water Issues*, by James E. McCarthy and Mary Tiemann. For information on ethanol, see CRS Report RL3 3290, *Fuel Ethanol: Background and Public Policy Issues*, by Brent D. Yacobucci.

[13] As classified under the old 1-hour ozone standard that was replaced with a new, 8-hour standard in 2004.

[14] U.S. Environmental Protection Agency (EPA), Office of Transportation Air Quality (OTAQ), Staff White Paper: Study of Unique Gasoline Blends ("Boutique Fuels"), Effects on Fuel Supply and Distribution and Potential Improvements, October 2001.

[15] In the case of MTBE, this equates to roughly 11% by volume.

[16] Renewable Fuels Association, "New Jersey Bans MTBE," Ethanol Report, Issue #226, July 15, 2005.

[17] USDA estimates that in 2008/2009, that percentage will increase to 34%. U.S. Department of Agriculture, Economic Research Service, *Feed Outlook*, January 14, 2009.

[18] For a detailed comparison of the renewable fuels legislation, see CRS Report RL32865, *Renewable Fuels and MTBE: A Comparison of Provisions in the Energy Policy Act of 2005 (P.L. 109-58 and H.R. 6)*, by Brent D. Yacobucci, Mary Tiemann, and James E. McCarthy.

[19] While the original mandate in P.L. 109-5 8 covered only gasoline, the expanded mandate applies to all transportation fuels as well as heating oil.

[20] U.S. Environmental Protection Agency, *EPA Finalizes Regulations for the National Renewable Fuel Standard Program for 2010 and Beyond*, EPA-420-F-10-007, Washington, DC, February 2010, *http://www.epa.gov/otaq/* renewablefuels/420f10007.htm.

[21] Steven Mufson, "A Boost for Corn-Based Ethanol?," *The Washington Post*, February 4, 2010, p. A15.

[22] For more information on EPA's waiver authority, see CRS Report RS22870, *Waiver Authority Under the Renewable Fuel Standard (RFS)*, by Brent D. Yacobucci.

[23] U.S. Environmental Protection Agency, *Regulation of Fuels and Fuel Additives: Changes to Renewable Fuel Standard; Final Rule*, EPA-HQ-OAR-2005-0161, Washington, DC, February 3, 2010, p. 14, http://www.epa.gov/otaq/ renewablefuels/rfs2-preamble.pdf.

[24] For more information on cellulosic biofuels, see CRS Report RL34738, *Cellulosic Biofuels: Analysis of Policy Issues for Congress*, by Kelsi Bracmort et al.

[25] A mixture of hydrogen and carbon monoxide that can be used to produce a variety of chemicals and fuels.

[26] On December 15, 2009, the Department of Energy finalized a rule establishing the incentive program. U.S. Department of Energy, "Final Rule: Production Incentives for Cellulosic Biofuels; Reverse Auction Procedures and Standards," 74 *Federal Register* 52867-52873, October 15, 2009.

[27] For more information on ethanol imports from CBI countries, see CRS Report RS21930, *Ethanol Imports and the Caribbean Basin Initiative (CBI)*, by Brent D. Yacobucci.

[28] For more information on drawbacks, see U.S. Customs Service, Drawback: A Refund for Certain Exports, Washington, February 2002.

[29] 19 U.S.C. 1313(p).

[30] Peter Rhode, "Senate Finance May Take Up Drawback Loophole As Part Of Energy Bill," EnergyWashington Week, April 18, 2007.

[31] For purposes of compliance with EPAct 1992, a covered vehicle fleet is one operated by an agency or company in a metropolitan area with at least 20 passenger vehicles in one location.

[32] For more information on vehicle purchase requirements, see the Federal Energy Management Program's Fleet Management program at
http://www1.eere.energy

[33] Center for Biological Diversity v. Abraham, N.D. Cal., No. CV-00027.

[34] U.S. Department of Energy, Office of Energy Efficiency and Renewable Energy, *Federal Fleet Compliance with EPA CT and E.O. 13149/E.O. 13423*, Fiscal Years 2000 through 2007.

[35] Light-duty diesel vehicles that meet specified emissions standards.

[36] 68 *Federal Register* 10319.

[37] For more information on vehicle tax incentives, see CRS Report RS22351, *Tax Incentives for Alternative Fuel and Advanced Technology Vehicles*, by Brent D. Yacobucci.

[38] For more information on the hybrid vehicle tax credit, see Internal Revenue Service, *Summary of the Credit for Qualified Hybrid Vehicles*, at *http://www.irs.gov/newsroom/article/0,,id=157632,00.html*, updated October 3, 2008.

[39] For example, depending on the technology used, processing coal into hydrogen could lead to significantly higher emissions of toxic compounds and carbon dioxide.

[40] Congress agreed to increase funding for hydrogen and fuel cell research from $185 million in FY2003 to $266 million in FY2004, $305 million in FY2005, $335 million in FY2006, $335 million for FY2007, and approximately $400 million for FY2008.

[41] For more information on the Bush Administration's initiatives, see CRS Report RS21442, *Hydrogen and Fuel Cell Vehicle R&D: FreedomCAR and the President's Hydrogen Fuel Initiative*, by Brent D. Yacobucci.

In: Cars, Climate and the EPA: Issues in… ISBN: 978-1-61728-963-7
Editors: Carolynne D. Burgdoerfer

Chapter 5

VEHICLE FUEL ECONOMY: NHTSA AND EPA'S PARTNERSHIP FOR SETTING FUEL ECONOMY AND GREENHOUSE GAS EMISSIONS STANDARDS IMPROVED ANALYSIS AND SHOULD BE MAINTAINED

United States Government Accountability Office

WHY GAO DID THIS STUDY

In May 2009, the U.S. administration announced plans to increase the Department of Transportation's (DOT) National Highway Traffic Safety Administration's (NHTSA) corporate average fuel economy (CAFE) standards and establish the Environmental Protection Agency's (EPA) greenhouse gas (GHG) emissions standards for vehicles. NHTSA redesigned CAFÉ standards for light trucks for model years 2008 through 2011, and some experts raised questions about the rigor of the computer modeling NHTSA used to develop these standards.

GAO was asked to review (1) the design of NHTSA and EPA's proposed standards; (2) how they are collaborating to set these standards; (3) improvements compared to a previous rulemaking, if any, NHTSA made to the modeling; and (4) the extent to which NHTSA analyzed the effects of past light truck standards and the accuracy of data used to set them.

GAO reviewed relevant rulemaking and modeling documents, and interviewed agency officials and other experts.

What GAO Recommends

GAO is recommending NHTSA and EPA document their collaborative process, formalize this relationship for the future, and conduct additional research and analyses of past light truck standards.

EPA agreed and DOT generally agreed with our recommendations.

What GAO Found

NHTSA and EPA have worked to propose CAFE and GHG standards that are generally aligned so manufacturers can build a single fleet of vehicles to comply with both. The standards are based on vehicle size and will cover model years 2012 to 2016. However, differences between the standards still exist because of variation in the legal authorities of each agency. For example, certain flexibility mechanisms designed to reduce compliance costs for manufacturers apply only to GHG standards, which could make aligning them with CAFE standards more difficult. However, potentially stricter penalties for GHG standard noncompliance could improve compliance with CAFE standards. Also, while NHTSA and EPA expect benefits from adopting a standard based on vehicle size, neither standard has a mechanism to ensure that a specific national target will be met.

NHTSA and EPA are collaborating by sharing resources and expertise to jointly set CAFE and GHG standards. From fiscal years 1996 through 2001, NHTSA was barred from using appropriated funds to raise CAFE standards. In contrast, EPA has continually expanded its automotive engineering expertise, including at its vehicle testing lab. As a result, EPA was able to contribute several original research studies to the proposed joint standards. Because this collaboration is not formally required and the agencies are not documenting the processes used—a recognized best practice—they may not be able to replicate them in the future.

To set the proposed standards, NHTSA improved upon the computer model compared to the version used that had been used to set the CAFE standards for 2008 through 2011 light trucks. One improvement was that

NHTSA increased the model's transparency by using publicly available, rather than confidential, data to develop a baseline fleet of vehicles. With EPA's input, NHTSA updated several data inputs such as technology costs and the cost of emissions. While experts GAO interviewed had varying critiques of NHTSA's model, there was no consensus on how NHTSA could further improve it. In particular, experts' opinions differed sharply on two studies, which reported opposing findings concerning the relationship between vehicle weight (a key factor in determining fuel consumption) and safety—suggesting that additional research may be warranted.

In part due to resource and data constraints, NHTSA has not yet evaluated its 2008 through 2011 light truck CAFE standards, which have a similar design to the new standards. Retrospective analyses of efforts and data inputs could inform NHTSA on the extent to which the standards met goals and provide means to improve the process of setting standards. Lacking such analysis, NHTSA does not know whether goals of the standards have been met or if changes are needed to the program. NHTSA officials said that while they would like to conduct such analyses, limited resources and time have prevented them from doing so, and they have no definitive plans to conduct them in the future.

ABBREVIATIONS

CAA	Clean Air Act
CAFE	corporate average fuel economy
CO_2	carbon dioxide
DOE	Department of Energy
DOT	U.S. Department of Transportation
EIA	Energy Information Administration
EISA	Energy Independence and Security Act of 2007
EPA	Environmental Protection Agency
EPGA	Energy Policy and Conservation Act
GHG	greenhouse gas
g/mi	grams per mile
mpg	miles per gallon
NAS	National Academy of Sciences
OMB	Office of Management and Budget
OMEGA	Optimization Model for Reducing Emissions of Greenhouse

Gases from Automobiles

February 25, 2010

The Honorable Edward J. Markey
Chairman
Subcommittee on Energy and Environment
Committee on Energy and Commerce House of Representatives

Dear Mr. Chairman:

In May 2009, the U.S. administration announced the National Fuel Efficiency Policy relating to cars and light trucks, which beginning in 2012, would not only increase corporate average fuel economy (CAFE) standards but also establish for the first time greenhouse gas (GHG) emissions standards. According to the administration, this effort will help to accomplish several goals. First, by helping to reduce oil consumption, CAFE standards could decrease the level of oil imports, in turn decreasing both the nation's economic vulnerability to oil price shocks and the trade deficit. Second, the administration intends for CAFE standards and GHG emissions standards to begin addressing global climate change by reducing emissions of GHGs such as carbon dioxide from the sector of the economy that has long been the fastest-growing source of these emissions—mobile sources like cars and trucks. Finally, this program represents a coordinated national approach to reducing GHG emissions and improving fuel economy, allowing auto makers to build a fleet of vehicles to meet one national standard rather than multiple standards set by federal and state governments.

Although the proposed standards offer potential benefits, they also impose costs. Given the nation's current economic challenges, it is imperative that in the course of setting new standards, agencies estimate as accurately as possible the benefits and costs the standards will impose on industry and consumers—as the standards are in part based on estimates of these costs and benefits. For example, if costs to consumers from increased standards are underestimated, then the standards might be too stringent, leading to high costs imposed on consumers without adequate benefits. The stringency of these standards—a key factor in generating costs to the industry and consumers—depends in part on analysis conducted by the National Highway Traffic Safety Administration (NHTSA), the agency responsible for regulating CAFE standards, and the Environmental Protection Agency (EPA), the agency responsible for regulating GHG emissions standards. Experts raised questions in 2006 when NHTSA set new CAFE standards for model year 2008 through 2011 light trucks. Specifically, experts expressed concerns about some of the data in the model NHTSA used to estimate the potential impact of increasing these standards. In response, NHTSA made changes to the model in preparation for establishing new standards for model years 2012 through 2016.

You asked us to review NHTSA and EPA's joint effort to set CAFE and GHG emissions standards. Specifically, we reviewed (1) the design of the proposed CAFE and GHG emissions standards, including similarities and differences between the two; (2) how NHTSA and EPA are collaborating in setting CAFE and GHG emissions standards and how the resources of both agencies are being used; (3) improvements compared to previous rulemakings, if any, made to NHTSA's process for setting standards—in particular, its regulatory impact analysis (the "Volpe model")—and for obtaining and validating data used in this model; and (4) the extent to which NHTSA analyzed the effects of its light truck standards for model years 2008 through 2011, as well as the accuracy of key data it used to establish these standards. To describe the design of the proposed CAFE and GHG emissions standards, we analyzed (1) rulemaking documents with information on the structure of the standards and how NHTSA and EPA aligned them and (2) legislation governing CAFE and GHG standards, as well as associated penalties for noncompliance. To describe how NHTSA and EPA are collaborating to set CAFE and GHG emissions standards, we reviewed and analyzed relevant rulemaking documents and legislation and interviewed NHTSA and EPA officials on their communication and coordination, analyzing this information against GAO criteria for evaluating communication and coordination among federal agencies. To identify improvements made to the Volpe model, we evaluated (1) documentation about the model against GAO criteria for developing cost estimates and assessing data reliability and (2) federal guidance for conducting regulatory and economic analyses. We interviewed experts and stakeholders with relevant expertise in areas such as economic modeling and automotive technology costs about data inputs and the design of the model. We also interviewed automobile industry stakeholders—including domestic and international automobile manufacturers and an association representing original equipment suppliers. Finally, to determine the steps NHTSA has taken to analyze the effects of the model year 2008 through 2011 light truck standards, we reviewed documentation related to these standards and interviewed NHTSA officials to determine whether NHTSA took steps to assess the outcomes of these standards or the accuracy of data it used to set these standards.

We conducted this performance audit from June 2009 through February 2010, in accordance with generally accepted government auditing standards. Those standards require that we plan and perform the audit to obtain sufficient, appropriate evidence to provide a reasonable basis for our findings and conclusions based on our audit objectives. We believe that the evidence obtained provides a reasonable basis for our findings and conclusions based on our audit objectives. (For more information on our scope and methodology, see app. I.)

BACKGROUND

The Energy Policy and Conservation Act (EPCA)[1], enacted in 1975, established CAFE standards with the goal of reducing oil consumption. EPCA required manufacturers to meet a single fleetwide CAFE standard for all cars and either a single standard or class standards for light trucks. The act provided the U.S. Department of Transportation (DOT) with the authority to administer the CAFE program, and DOT delegated that authority to NHTSA. In addition, other federal agencies have played a role in the CAFE program (see table 1). For example, under EPCA, EPA is responsible for the development of CAFE testing and calculation procedures.[2] When it was enacted, EPCA specified that the standard for passenger cars would be 18 miles per gallon (mpg) in 1978, rising to 27.5 mpg by 1985, but it permitted NHTSA to determine the standard for light trucks through rulemakings. As required in EPCA, NHTSA began setting CAFE standards for light trucks at the "maximum feasible level" and made incremental increases to these standards from 1979 through 1996. During that time, the light truck CAFE standard increased from 17.9 mpg to 20.7 mpg. However, from fiscal years 1996 through 2001, NHTSA was barred from using appropriated funds made available in DOT's appropriation to raise CAFE standards.[3] The CAFE standard for cars remained at the 1985 setting of 27.5 mpg through model year 2010. The first increase in CAFE standards for cars since 1985 will take place for model year 2011 cars.

After years of little CAFE-related activity or movement in the two standards, several changes took place. According to NHTSA officials, DOT requested that the appropriations ban be lifted so that they could raise CAFE standards. The ban was lifted beginning in fiscal year 2002, and in 2003, NHTSA promulgated increased CAFE standards for light trucks for model years 2005 to 2007. In 2006, NHTSA issued another rule to increase and reform the standards for light trucks, which we refer to as the model year 2008 through 2011 light truck standards. In this light truck rulemaking, NHTSA transitioned from a single CAFE standard applicable to each manufacturer's fleet to a reformed, attribute-based standard based on a vehicle's "footprint," or the size of its wheelbase multiplied by its average track width. The move from a single standard for all light trucks to attribute-based standards for each light truck vehicle model based on a vehicle's footprint was designed to address a number of downsides to "unreformed"[4] CAFE standards, including potential safety implications and consumer choice limitations. The Energy Independence and Security Act of 2007 (EISA)[5] amended EPCA to require not only light truck but also passenger car standards to be based on an attribute-based curve and for the fuel economy of the entire industrywide fleet—including cars and light trucks—to reach an average of 35 mpg by model year 2020. Subsequent to the enactment of EISA, in 2008, NHTSA proposed CAFE standards based on vehicle footprints for passenger cars and light trucks for model years 2011 through 2015.[6] However, a final rule was issued only for model year 2011 standards in March 2009[7]—a rulemaking effort and CAFE standard that we refer to as the model year 2011 CAFE standard. The goal of this final rule was to reach an estimated fleet average—or target—of 30.2 mpg for cars and 24.1 mpg for light trucks in model year 2011.

In recent years, public concerns have grown about the relationship between the emission of GHGs and global climate change. According to the Intergovernmental Panel on Climate Change—a United Nations organization—global atmospheric concentrations of GHGs have increased as a result of human activities, contributing to a warming of the earth's climate. If unchecked, this could have serious negative effects, such as rising sea levels and coastal flooding worldwide.

Automobiles represent a significant share of GHG emissions. According to EPA, in 2007, personal vehicle use accounted for 17 percent of total GHG emissions in the U.S. In 2007, the United States Supreme Court ruled that EPA has the statutory authority to regulate GHG emissions from new motor vehicles under the Clean Air Act (CAA) because greenhouse gases meet the CAA's definition of an air pollutant. Furthermore, the Supreme Court held that EPA must regulate GHGs as such if EPA finds them to be an endangerment to public health or welfare.[8] Subsequent to this decision, EPA issued a final

Endangerment Finding of GHG emissions[9] in December 2009, laying the foundation for setting GHG emissions standards for vehicles.[10]

Table 1. Federal Agency Roles in CAFE

Agency	Role
NHTSA	• Set and enforce CAFE standards
EPA	• Consultant to NHTSA in setting CAFE standards • Conduct vehicle testing to determine manufacturer fuel-economy levels, which are provided to NHTSA and individual manufacturers • Function as lead in setting GHG emissions standards and joint partner with NHTSA in rulemaking for proposed CAFE and GHG standards
DOE	• Consultant to NHTSA in setting CAFE standards

Source: GAO analysis.

In addition, in 2005, citing compelling and extraordinary impacts of climate change on the state, California filed a request with EPA for a waiver of CAA preemption[11] to set GHG emissions standards for new motor vehicles starting in the 2009 model year. The CAA directs EPA to grant a waiver unless EPA finds (1) the state's protectiveness determination was arbitrary and capricious, (2) the state's standards are not needed to meet "compelling and extraordinary conditions," or (3) the state's standards are inconsistent with section 202(a) of the CAA (provisions related to technical feasibility and lead time to manufacturers). Under certain conditions set forth in the CAA, other states may adopt California's motor vehicle emissions standards. The automobile industry brought litigation in several states, including California, alleging, among other claims, that the state standards were preempted by EPCA (which preempts state standards relating to fuel economy). Federal district courts in Vermont and California ruled against such claims, in the only two cases to be judged on their merits to date.[12] California's waiver request was initially denied by the prior administration. EPA determined that California's standards were not needed to meet compelling and extraordinary conditions, as required by the CAA, because global climate change and local or regional factors represent different causal links affecting air pollution in California—and previous waivers have addressed only the local or regional air pollution problems. In addition, EPA found that the effects of climate change in California are not compelling and extraordinary when compared to the rest of the country.[13] GAO found in January 2009[14] that the "compelling and extraordinary" test had never before been used to completely deny a waiver request. The current administration also found that the denial was a historical anomaly, reconsidered the request, and granted the waiver in June 2009 after finding that it should not have been denied under any of the statutory factors.[15] Petition for review of this decision filed by the U.S. Chamber of Commerce and the National Automobile Dealers Association is now pending in front of the U.S. Court of Appeals for the District of Columbia Circuit.[16]

In response to the EISA's call for higher CAFE standards and California and other states' desire to establish fuel economy or GHG emissions standards, the current administration announced its National Fuel Efficiency Policy in May 2009. This policy involves setting higher CAFE standards for model years 2012 through 2016 for cars and light trucks, as well as new GHG emissions standards by EPA during this same period. As a result, NHTSA and EPA are conducting a joint rulemaking to increase CAFE standards and set new GHG emissions standards. [17] (See figure 1 for a timeline of major CAFE and GHG emissions standards milestones.) before been used to completely deny a waiver request. The current administration also found that the denial was a historical anomaly, reconsidered the request, and granted the waiver in June 2009 after finding that it should not have been denied under any of the statutory factors.[15] Petition for review of this decision filed by the U.S. Chamber of Commerce and the National Automobile Dealers Association is now pending in front of the U.S. Court of Appeals for the District of Columbia Circuit.[16]

In response to the EISA's call for higher CAFE standards and California and other states' desire to establish fuel economy or GHG emissions standards, the current administration announced its National Fuel Efficiency Policy in May 2009. This policy involves setting higher CAFE standards for model years 2012 through 2016 for cars and light trucks, as well as new GHG emissions standards by EPA during this same period. As a result, NHTSA and EPA are conducting a joint rulemaking to increase CAFE standards and set new GHG emissions standards. [17] (See figure 1 for a timeline of major CAFE and GHG emissions standards milestones.)

The proposed joint rule would increase CAFE standards to achieve an estimated fleetwide average of 34.1 mpg and implement GHG emissions standards to achieve an estimated fleetwide average of 250 grams per mile (g/mi) of carbon dioxide (CO_2) by model year 2016. The agencies jointly issued a Notice of Upcoming Joint Rulemaking in May 2009, issued a Proposed Rulemaking and held three public hearings across the country in

September 2009, held a 60-day public comment period that ended in November 2009, and plan to issue the final rules by April 1, 2010.[18] (Figure 2 shows the changes to CAFE standards over time, including the proposed standards).

1978	The corporate average fuel economy (CAFE) standards for newly manufactured passenger cars—established in 1975 by the Energy Policy and Conservation Act (EPCA)—take effect. The standards were initially set at 18 mpg.
1979	The first set of light truck CAFE standards, which EPCA directed NHTSA to establish administratively, take effect. The light truck standards vary initially depending on type of vehicle (2-wheel vs. 4-wheel drive) and increase incrementally each year until 1996.
1985	The final incremental increase in car CAFE standards under EPCA takes effect, and newly manufactured cars were required to average 27.5 mpg.
1986	In response to petitions from automakers who noted that consumers were demanding larger cars and engines, largely due to a decline in gasoline prices, NHTSA reduces the CAFE standard for passenger cars to 26.0 mpg.
1990	NHTSA returns CAFE standards for passenger cars to 27.5 mpg.
1992	NHTSA replaces the separate standards for 2-wheel vs. 4-wheel drive light trucks with a single standard of 20.2 mpg.
1996	NHTSA's appropriation prohibits it from conducting any work on CAFE, which freezes the standard for light trucks at 20.7 mpg.
2001	Congress lifts its prohibition on NHTSA conducting work related to CAFE.
2003	NHTSA issues a rule for light truck CAFE standards for model years 2005 to 2007.
2004	The California Air Resources Board (CARB) adopts the nation's first GHG rule, which requires automakers to begin selling vehicles in California with incrementally decreasing GHG emission levels between 2009 and 2016.
2005	CARB submits a request for a waiver of Clean Air Act (CAA) preemption for its GHG emission standards for motor vehicles to take effect in 2009.
2006	NHTSA issued final rule reforming light truck standards for model years 2008 to 2011 and requested Congress to provide authority to reform passenger car standards.

2007 (April)	The U.S. Supreme Court rules that greenhouse gases (GHG) meet the CAA definition of an air pollutant and that EPA has the statutory authority to regulate these emissions from new motor vehicles under the act.
2007 (December)	Congress enacts the Energy Independence and Security Act (EISA), which reforms car standards and calls for CAFE standards to reach an industrywide fleet average (i.e., all manufactured vehicles) of 35 mpg by 2020.
2008	NHTSA's reformed light truck standards, which are designed around the "footprint" (or wheelbase multiplied by track width) of a vehicle rather than as a single standard, take effect.
2009 (March)	NHTSA finalizes CAFE standards for cars and light trucks, to be implemented in 2011, increasing car standards for the first time in about 20 years.
2009 (May)	The current administration announces plans for new CAFE standards, beginning in 2012 and increasing to a fleet average of 35.5 miles per gallon by 2016. The new CAFE standards are to be harmonized with the new vehicle GHG emissions standards to be set by EPA and are based on vehicle footprint for both passenger cars for the first time, as well as light trucks.
2009 (July)	After initially being denied, California is granted a waiver of CAA preemption by EPA, giving the state authority to set GHG emission standards for vehicles in the future. California elects to adopt the national standard being developed by EPA, but also begins looking ahead to standards in 2017.

Source: GAO.

Figure 1. Timeline of Major CAFE and GHG Emissions Standards Milestones

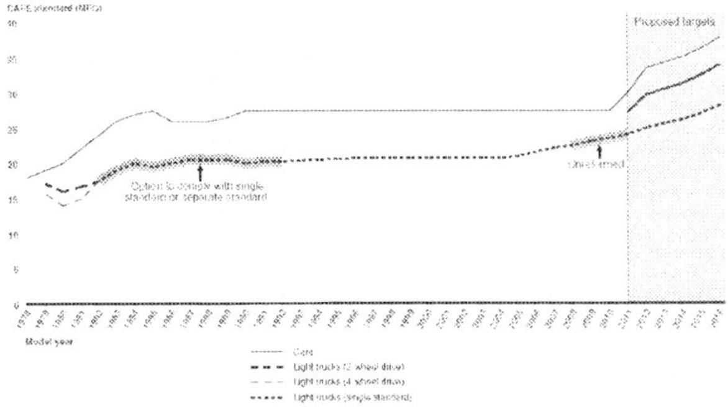

Source: NHTSA.

Figure 2. Historical Changes to CAFE Standards and Proposed Future Targets

In the proposed rule, NHTSA and EPA estimate that the proposed standards will result in both benefits and costs:[19]

- *Potential benefits for consumers and society.* The agencies estimate that the new standards will result in approximately 1.8 billion barrels of oil savings and 950 million metric tons of carbon dioxide emissions reductions over the lifetime of vehicles sold in model years 2012 through 2016. In addition, the agencies estimate that new and more fuel-efficient vehicles will save consumers more than $4,000 in gasoline costs over a model year 2016 vehicle's lifetime.

- *Potential costs for consumers, automobile manufacturers, and others.* The agencies estimate that the proposed standards would require manufacturers to incorporate additional fuel-saving technology into vehicles, which would increase the average cost of a model year 2016 vehicle by around $1,100. As a result, this will increase the purchase price of vehicles for consumers, or manufacturers will receive lower profits from vehicle sales, or both. However, the agencies estimate that the total benefits of the proposed standards will outweigh the costs, providing net benefits to society of nearly $200 billion over the lifetimes of the model year 2012 to 2016 vehicles. In addition, the estimated lifetime fuel savings exceeds the $1,100 increase in vehicle cost for a model year 2016 vehicle, yielding a net savings of about $3,000 for consumers.

Although NHTSA and EPA Worked to Propose CAFE and GHG Emissions Standards That Are Aligned, the Programs Have Several Key Differences

Although the Proposed Standards Based on Vehicle Footprint Should Result in Benefits, Actual Vehicle Sales May Affect the Level of Benefits Realized

Although the proposed CAFE and GHG emissions standards are distinct and automobile manufacturers will be subject to both sets, EPA and NHTSA have worked to develop standards that are aligned (what the agencies refer to as "harmonized") with the intention that manufacturers can build one fleet of vehicles to comply with both sets of standards. This should lower the cost of compliance for manufacturers compared to a case in which the standards were set separately and without regard for the other's design. This harmonization is possible because fuel economy and GHG emissions have a clear and direct relationship—specifically, vehicle tailpipe carbon dioxide emissions are directly related to the quantity of fuel burned.[20] Given the relationship between GHG emissions and fuel economy, actions to increase fuel economy also necessarily reduce GHG emissions; therefore, manufacturers can use the same technologies to help meet both standards.

NHTSA and EPA have proposed standards for both passenger cars and light trucks that are based on vehicle footprint so that each vehicle is subject to a target level based on its footprint, with smaller vehicles having a stricter target (see figure 3). The footprint-based standard is applied to individual vehicle models based on the size of each vehicle. Because each manufacturer sells a different mix of vehicle sizes, under the proposed standards each manufacturer will have different CAFE and GHG emissions standards.

NHTSA first adopted a footprint-based approach—as opposed to a single fleetwide standard—for model year 2008 through 2011 light truck standards.[21] A number of the experts we interviewed supported the current approach of subjecting both passenger car and light truck fleets to footprint-based standards. In the model year 2008 through 2011 light truck rule, NHTSA cited several potential benefits of a footprint-based approach over a single, fleetwide CAFE standard, including the following:

- *Larger reductions in oil consumption.* Oil consumption would be reduced because automakers would be required to improve the fuel economy of vehicles of all sizes rather than only those near the standard.

- *Enhanced safety.* Manufacturers would not have an incentive to comply with CAFE standards by pursuing strategies that compromise safety— such as (1) reducing the size of vehicles (applicable fuel-economy targets now become higher as size decreases) or (2) designing models to be classified as light trucks rather than cars, which can increase a vehicle's propensity to roll over—in order to comply with CAFE standards. Under a single standard, manufacturers could reduce vehicle size as one approach for CAFE compliance.

- *More even disbursement of the regulatory cost burden.* Fuel-economy improvements would be spread across the industry, instead of concentrating on manufacturers of heavier, lower fuel-economy vehicles.

- *Addressing concerns about consumer choice.* Manufacturers now must improve the fuel economy of all light trucks, regardless of size, which addresses criticisms that single, fleetwide CAFE standards were hindering the efforts of some companies to offer a mix of vehicles matching consumer desires. For instance, under the previous system, instead of installing more fuel-saving technologies across their fleets, manufacturers might have moved toward building fewer large vehicles and more small vehicles to meet new CAFE standards, even though consumers typically have not demanded small vehicles. In a footprint-based standard, manufacturers must improve the fuel economy of all light trucks, no matter their size.

The CAFE requirement for each manufacturer—which is the basis for determining compliance[22]—will be determined at the end of the model year based on actual production. For example, manufacturers selling a greater proportion of large vehicles will have a lower average target to meet than will manufacturers focusing on smaller vehicles. Based on estimated sales projections, the proposed targets are estimated to achieve an average of 34.1 mpg across all model year 2016 vehicles sold.

While NHTSA and EPA expect benefits from adopting a standard based on vehicle footprint and predict that the administration's goal of a fleetwide average 34.1 mpg and 250 grams per mile carbon dioxide in 2016 will be met, there is no guarantee that a specific national target will be achieved.[23] This is a tradeoff of adopting a footprint standard compared to the single national CAFE standard NHTSA used in the past. Because the actual fleetwide fuel-economy levels will depend on actual vehicle sales— specifically, the size of cars consumers buy—there is the possibility that the actual fleetwide mpg in 2016 will be higher or lower and realized costs and benefits of the standards will be higher or lower than estimated.

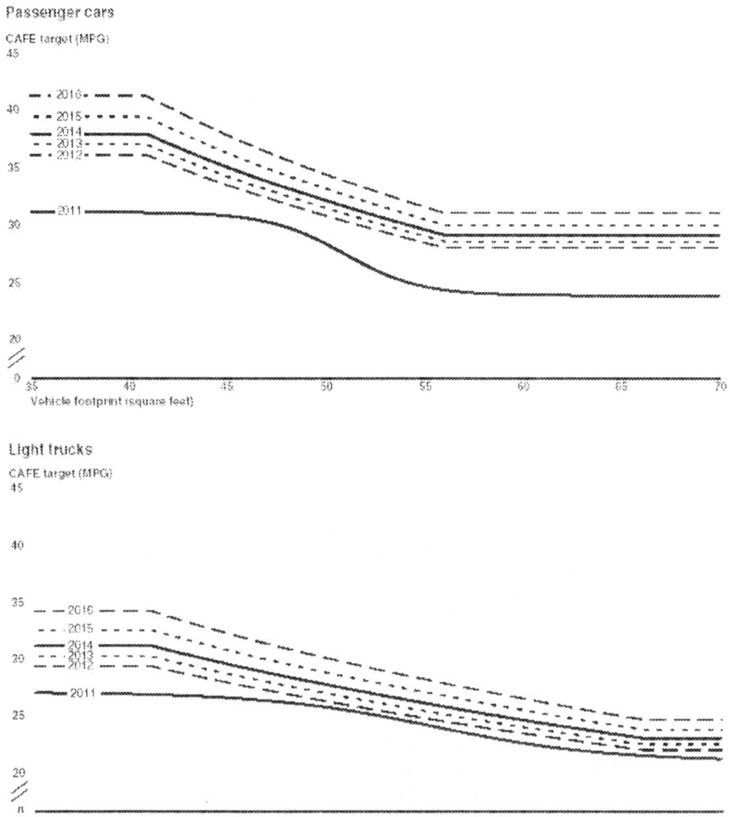

Source: Notice of Proposed Rulemaking for CAFE standards for MY 2012 to 2016.

Figure 3. Proposed CAFE Footprint Curves for Passenger Cars and Light Trucks, Model Years 2012 through 2016 and Existing 2011 Curve

For example, even though all of the vehicles in each manufacturer's fleet may be in compliance with its footprint-based requirement, manufacturers may sell a greater number of large-footprint vehicles than predicted, which would lower each manufacturer's CAFE requirement. If this is the case, the national fleet may not reach the target of 34.1 mpg by 2016, and the estimated benefits of the standards, which assume achieving a national fleetwide average of 34.1 mpg, would not be fully realized.[24] The opposite, however, could also be the case. If a greater number of smaller vehicles (generally with higher CAFE levels) are sold than expected, manufacturers will have higher CAFE requirements, the national fleet may exceed the target of 34.1 mpg, and estimated benefits assuming a fleetwide average of 34.1 mpg would be exceeded (see figure 4). Similar scenarios could occur with respect to EPA's GHG standards.

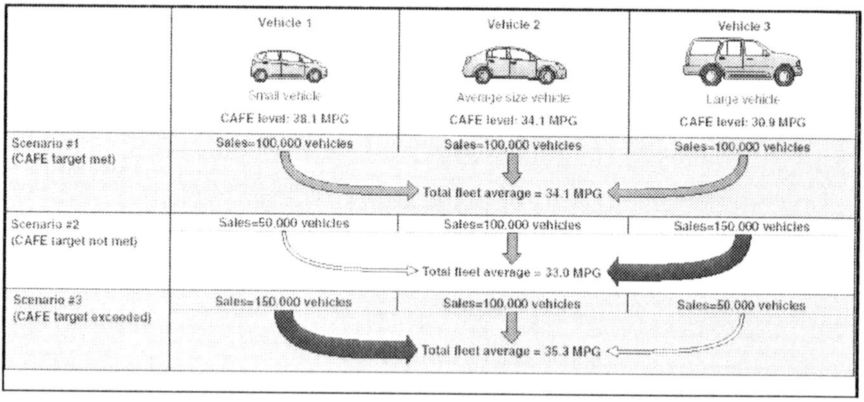

Source: GAO analysis of proposed CAFE standards.

Figure 4. Potential Scenarios for Meeting CAFE Targets, Based on Varying Vehicle Sales

Variation in the Standards, Which Result Primarily from Differences in Legal Authorities, May Present Challenges, but GHG Penalties May Increase Compliance

Several key differences between the EPA and NHTSA standards largely arise from the legal authorities under which the standards are set. NHTSA's authority to administer the CAFE program is derived from EPCA, as amended by EISA, requires that NHTSA, for passenger cars and light trucks in each future model year, establish standards at "the maximum feasible average fuel-economy level that it decides manufacturers can achieve in that model year." EPCA further directs NHTSA to make this determination based on consideration of four statutory factors: technological feasibility, economic practicability, the effect of other standards of the government on fuel economy, and the need of the nation to conserve energy. However, the law does not direct NHTSA on how to balance these four factors—which can conflict—thereby giving NHTSA discretion to define, give weight to, and balance the four factors based on the circumstances in each CAFE rulemaking. Furthermore, how NHTSA balances these four factors can vary from rulemaking to rulemaking. For example, in the model year 2012 through 2016 rulemaking, NHTSA cited economic practicability concerns—given the state of the economy and the financial state of automakers—to set standards at a level lower than it otherwise could have in accordance with Office of Management and Budget (OMB) guidelines on federal regulatory impact analysis.[25] In addition to the four statutory factors, NHTSA also considers the potential for adverse safety consequences and consumer demand when establishing CAFE standards.

EPA's authority to set GHG standards is derived from the CAA, which authorizes EPA to regulate emissions of air pollutants from all mobile source categories. EPA must prescribe standards for the emission of any air pollutant from motor vehicles which causes or contributes to air pollution that endangers public health or welfare. In prescribing these statutory standards, EPA considers such issues as technology effectiveness, cost of compliance, the lead time necessary to implement the technology, safety, energy impacts associated with the use of the technology, and other impacts on consumers. EPA has the discretion to consider and weigh these various factors, particularly those related to issues of technical feasibility and lead time.

Some differences affect the process each agency must use to set standards, which in turn leads to key differences between the standards. For example, EPCA requires that EPA, in testing fuel economy of passenger vehicles, use 1975 test procedures or procedures that give comparable results under which air conditioning is not turned on. As a result, manufacturers cannot realize the benefits of air conditioning improvements for complying with CAFE standards, and NHTSA has, to date, not taken into account air conditioning improvements when setting CAFE standards. [26] Under the CAA, however, EPA is not subject to the same limitations, and its proposed GHG standards account for air conditioner improvements. Specifically, the mpg equivalent of EPA's 2016 target of 250 g/mi of CO_2 emissions corresponds to 35.5 mpg. The CAFE target is 34.1 mpg because it cannot account for air conditioning improvements.

In addition, certain flexibility mechanisms designed to achieve and reduce the cost of compliance are authorized by one program but not the other.

This creates potential challenges to harmonization and for manufacturers attempting to manage the design of a fleet. For example, EPA's proposed GHG standards offer a "temporary lead time" mechanism for manufacturers that sell a limited number of vehicles in the U. S.[27] Although this specific flexibility does not exist in the CAFE standards, under EPCA, NHTSA may exempt qualifying small-volume manufacturers (defined as manufacturers that produce under 10,000 vehicles worldwide annually) from the passenger car standard for a model year. As a result, manufacturers that are able to take advantage of EPA's temporary lead time mechanism to comply with GHG standards may face challenges in complying with CAFE standards. Some experts we met with said that these inconsistencies in flexibility mechanisms between the two sets of standards may present challenges to some manufacturers in meeting the harmonized standards.

Mechanisms available for enforcing the standards also differ between the two agencies due to statutory differences. For example, the Clean Air Act prohibits the sale of vehicles without a certificate of conformity from EPA which indicates that the vehicle meets applicable emission standards.[28] If EPA determines that a vehicle does not meet the emission standards, it may not issue a certificate, thus preventing the manufacturer from legally selling the vehicle. The Clean Air Act also gives EPA authority to recall noncompliant vehicles. NHTSA can take neither of these actions. Because a CAFE standard applies to a manufacturer's entire fleet for a model year, CAFE fines are assessed for the entire noncomplying fleet. Pursuant to EPCA, fines associated with CAFE noncompliance are currently $5.50 for every tenth of an mpg a manufacturer's fuel economy is short of the standard multiplied by the number of vehicles in a manufacturer's fleet for a given model year. NHTSA recognizes that some manufacturers regularly pay fines instead of complying with CAFE standards; in particular, many European manufacturers pay fines each year. Fines for CAFE standards have not been increased since 1997, and GAO has reported that, as a result, CAFE penalties may not provide a strong enough incentive for manufacturers to comply with CAFE. NHTSA officials noted that under EPCA, NHTSA has the authority to raise the fines up to $10 per tenth of an mpg. However, raising fines requires an analysis finding that substantial energy conservation would result and that raising fines would not have substantially deleterious impact on the U.S. economy. GAO has recommended that agencies collecting penalties regularly conduct these types of analyses.[29]

In contrast to CAFE fines, penalties for violation of a motor vehicle emission standard under the CAA, which may be much higher, are determined on a per-vehicle basis. The CAA gives EPA broad authority to levy fines and require manufacturers to remedy vehicles if the agency determines there are a substantial number of noncomplying vehicles.[30] EPA must consider an assortment of factors, such as the gravity of the violation, the economic impact of the violation, the violator's history of compliance, and other matters,[31] in determining the appropriate penalty. The CAA does not authorize manufacturers to intentionally pay fines as an alternative to compliance, and

EPA does not include in its standard-setting modeling analysis the option for manufacturers to pay fines instead of compliance. Manufacturers may be subject to fines as high as $37,500 per vehicle under Section 205 of the CAA. Given that fines for noncompliance with GHG standards may be higher than fines for noncompliance with CAFE, having harmonized standards may provide incentives to manufacturers that have traditionally chosen to pay CAFE penalties instead of complying with standards, to comply with both sets of standards.

Although the Agencies Closely Collaborated and Capitalized on EPA's Recent Research in Setting Standards, Joint Rulemaking for Future Standards Is Not Guaranteed

EPA and NHTSA Are Collaborating through a Joint Rulemaking Process, Which Represents an Expansion of EPA's Role Compared to Previous CAFE Rulemakings

In conducting the joint rulemaking, the agencies have collaborated on major tasks. For example, the two agencies coordinated time frames so that key milestones of each rulemaking—such as issuance of the Proposed Rulemaking and time frames for public comment—happened at the same time. This enabled manufacturers to learn about both new standards at the same time and plan appropriately. Officials of both agencies told us that staff from both agencies met on a regular basis, often daily, to coordinate their efforts throughout the rulemaking process. In addition, according to agency officials, the two agencies formed a number of joint technical teams to examine data used in modeling efforts—for instance, one team examined data on automotive technology that can improve fuel economy and reduce GHG emissions—to ensure that both agencies were using similar data and making similar assumptions to develop standards. As a result of these efforts, each agency had significant input into the development of both sets of standards.

EISA mandated NHTSA to consult with both EPA and the Department of Energy (DOE) in prescribing CAFE standards beginning with model year 2011.[32] NHTSA's use of EPA's expertise in environmental issues and DOE's expertise in energy efficiency in informing CAFE standards is important given CAFE's environmental and energy-security implications. For example,

NHTSA has prepared draft and final environmental impact statements, as required by the National Environmental Policy Act, discussing the environmental implications of recent CAFE rulemakings, and EPA has reviewed and provided input on that work. However, EPA's role in the joint CAFE and GHG emissions rulemaking goes beyond the EISA requirement for consultation. For example, EISA does not require either EPA or DOE to participate in CAFE rulemaking at as high a level as EPA has in the current joint CAFE and GHG emissions rulemaking.

This level of EPA involvement in the proposed 2012 through 2016 CAFE and GHG rulemaking is greater than EPA's involvement in previous CAFE rulemakings, particularly prior to NHTSA's proposal of CAFE standards for model year 2011. For the model year 2011 proposal, NHTSA and EPA staff jointly assessed which technologies would be available for those model years and their effectiveness and cost. They also jointly assessed key economic and other assumptions affecting the stringency of future standards. Finally, they worked together in updating and further improving the model that had been used to help determine the stringency of the model year 2008 through 2011 light truck standards. However, even in the rulemaking for model year 2011, EPA did not devote as many resources or have as much involvement in setting CAFE standards as it did in the model year 2012 through 2016 proposed CAFE and GHG rulemaking.

The increased involvement by EPA as an equal partner in the proposed model year 2012 through 2016 CAFE and GHG emissions rulemaking came at the direction of the current administration, when it announced plans to increase CAFE standards and introduce GHG emissions standards for vehicles. EPA officials noted that the involvement of the White House and clear directives to both the Secretary of Transportation and Administrator of EPA for a collaborative approach caused both agencies to commit to the joint process, which officials viewed as successful.

Both NHTSA and EPA Used Computer-Based Models to Conduct Analyses That Inform the Level of Standards; Results Were Largely Similar

To determine the appropriate level of CAFE and GHG emissions standards, NHTSA and EPA each conducted its own regulatory impact analysis using computer models. NHTSA used a model developed by the Volpe National Transportation Systems Center (referred to as the Volpe model), earlier versions of which have been used in previous CAFE rulemakings. The model estimates the costs and benefits to manufacturers, consumers, and society of differing levels of CAFE standards. (See app. II for an in-depth description of NHTSA's Volpe model.) EPA developed a similar model called the Optimization Model for Reducing Emissions of Greenhouse Gases from Automobiles (OMEGA) to conduct a similar analysis of and inform its proposed GHG standards.

While the models are distinct from one another, and NHTSA and EPA each conducted its own modeling, the two agencies collaborated on and coordinated this work. In particular, the OMEGA model and Volpe model generally used consistent data inputs and assumptions—for example, the same economic assumptions and, to the extent possible given structural differences between the models, consistent data on vehicle fleets and fuel- saving technologies. According to officials from both agencies, the two agencies worked closely together to develop these data inputs and assumptions. NHTSA's and EPA's analyses are also structured similarly and have two components—one that attempts to determine manufacturer response to the standards and another that estimates the effects of the proposed standards on manufacturers, consumers, and society.

In addition, although the two models differ in several ways, analyses conducted with each model produced similar results, helping to validate each modeling effort. Some differences involve the treatment of compliance flexibilities or credits—mechanisms created in a standard to reduce the cost of compliance for manufacturers. Other differences involve how the models account for manufacturers conducting multiyear product planning and how technologies were carried over between model years. Both NHTSA and EPA conducted analyses of the respective effects of the proposed CAFE and GHG standards. However, despite differences between the two models, the aggregate results were largely similar.

Although They Collaborated Closely, the Agencies Provided Differing Levels of Research and Studies to Support Rulemaking

Although NHTSA contributed research to the rulemaking process, it faced challenges in doing so. NHTSA contributed research on fuel efficiency and costs. For example, NHTSA officials said that they conducted new research related to estimating the rebound effect[33] and the costs of oil imports. In 2008, during the development of the model year 2011 rule, NHTSA contracted with an automotive consulting firm to review comments from stakeholders during the public comment period of the rulemaking, which resulted in some technology costs being updated.[34] NHTSA officials said that this work helped improve its analysis. NHTSA also contributed safety research. However, NHTSA has not recently undertaken new safety research to support the current proposed standards,[35] despite significant and ongoing controversy over vehicle safety and CAFE standards, as well as changes in technology available to reduce vehicle weight.[36] According to NHTSA officials, NHTSA has made such research a priority for the near-future in order to support future CAFE rulemaking.

In addition, while NHTSA contracted with the National Academy of Sciences (NAS) to provide an updated report on the costs of fuel-saving technologies,[37] and NAS held its first public meeting for this work in September 2007, this work was not completed in time to support analysis for the Notice of Proposed Rulemaking. EISA mandated NHTSA to contract with NAS to receive updates to its earlier report of fuel-saving technology cost and effectiveness in 5-year intervals until 2025. We noted in previous work that both experts and NHTSA officials said it would be ideal to complete and update such work before NHTSA issues a new car or light truck fuel-economy standard.[38] Also, NAS work on technology costs in 2002 was generally viewed by a wide range of experts as being thorough and unbiased. While NAS indicated in a preliminary report that it would finish its work by spring 2008, according to NAS officials, they required more time to acquire technology cost data than initially anticipated. As a result, the final NAS study has not yet been published and was not available to inform analysis for EPA and NHTSA's September 2009 Notice of Proposed Rulemaking.

EPA contributed research in time to provide analysis for the proposed rule. It also contributed funding to a greater degree, especially when compared with past CAFE rulemakings where EPA's role was limited to consulting. For example, EPA conducted or contracted for three peer- reviewed studies to support the rulemaking and the modeling efforts. According to EPA officials, these studies included

- an ongoing $1.1 million study done in conjunction with a consulting firm to determine the direct manufacturing costs of fuel-saving and GHG emissions-reducing technologies—a key input in both agencies' models;[39]
- a $40,000 assessment of indirect costs of manufacturing more fuel-efficient vehicles;[40] and

- a $1 million vehicle simulation modeling study done in conjunction with a consulting firm to refine estimates of emissions reduction and fuel-economy improvements stemming from combinations of technology.[41]

These studies provided the analysis of both CAFE and GHG standards with updated information and data.

The difference in the extent of new research that NHTSA and EPA conducted for this rulemaking likely results from differences in resources available to the agencies in the recent past. As we mentioned previously, from fiscal years 1996 to 2001—about 6 years—NHTSA was prohibited from using appropriated funds to change CAFE standards. According to NHTSA, the agency lost staff with expertise in this area as a result and did not begin to hire additional automotive engineers until summer 2009. By comparison, EPA has been able to develop and maintain automotive engineering expertise. This expertise has proved helpful in setting GHG emissions standards for automobiles. For example, EPA has been home to the National Vehicle and Fuel Emissions Laboratory since 1971, and in the early 1990s, it expanded its activities to conduct research and development of technologies used to reduce emissions, which are often marketed and licensed to the automobile industry. Although NHTSA brings safety expertise to CAFE standards, which has been a concern with raising CAFE standards in the past, the agency's primary mission and expertise is in vehicle safety, not vehicle power train design and the impact of vehicle emissions on the environment. Thus NHTSA cannot be expected to have the same level of in-house expertise related to vehicle power train design and environmental issues as EPA.

Although the Collaboration between NHTSA and EPA Resulted in Improvements, the Process Is Not Being Formalized or Documented for Future Use

Although the agencies had to work quickly, the joint proposed model year 2012 through 2016 rulemaking has met all of its milestones to date, and the agencies stated that the collaboration has been successful. This is the first time NHTSA and EPA are conducting a joint rulemaking together. The agencies conducted the joint rulemaking under tight time frames and have met all key milestones, such as publishing information about the rule and receiving and responding to public comments. However, the fast pace has left little time or resources to document any effective or efficient processes so they could be used in the future. From the administration's May 2009 release of the Notice of Upcoming Joint Rulemaking to the expected release of the rule, less than 11 months will have transpired. By comparison, according to NHTSA officials, other recent CAFE rulemakings have taken a minimum of 14 months. The accelerated timeline in the current rulemaking stemmed in part from the statutory requirement that NHTSA issue new CAFE standards 18 months prior to the beginning of the model year that will be affected and from the current administration's announcement regarding the development of the new standards in May 2009. In order to issue harmonized standards at the same time, both EPA and NHTSA had to adhere to an accelerated timeline.

Despite the dual challenge of conducting a joint rulemaking for the first time and on a compressed timeline, some experts we spoke with thought that the two agencies worked well with each other and hoped they would continue to do so. In addition, both agencies found the collaborative partnership to be successful. The proposed standards cover model years 2012 through 2016, and while it is not clear how fuel economy and GHG emissions will be regulated after 2016, industry stakeholders and others have said that they would like NHTSA and EPA to begin working on the next set of standards in the near future. Officials with the California Air Resources Board said that the state is already considering state GHG emissions standards that would take effect in 2017, and depending on the stringency of federal standards at that time, California may opt to implement its own more stringent standards. Many industry stakeholders we interviewed said that they prefer a national program with harmonized standards over different federal and state standards because multiple standards could substantially increase compliance costs. Some expressed interest in EPA and NHTSA considering CAFE and GHG emissions standards for model years beyond 2016 as soon as possible in order to better ensure harmonized national standards and to give manufacturers appropriate lead time to meet standards.[42]

Although we found interest in NHTSA and EPA developing standards for model years beyond 2016, two issues could prevent the agencies from replicating this effort in the future:

- *The processes for coordinating the rulemaking have not been documented by either agency.* Documented processes that the two agencies would follow—detailing how each communicated, shared resources, and set plans—would help ensure that best practices are followed and that resources are used efficiently. As GAO has reported,[43] such guidance can aid regulatory programs by improving efficiency and ensure that benchmarks and time frames are met. In addition, by publishing such documentation, the agencies can increase the transparency of their programs and processes. However, the two agencies have not documented the processes for use during future rulemakings, and officials at both agencies report they currently have no plans to do so. EPA officials, however, told us that documenting the processes would be a worthwhile task.

- *The two agencies are not legally required to continue coordinating in setting CAFE and GHG emissions standards.* As noted, EISA mandated NHTSA to consult with EPA and DOE in setting CAFE standards beginning with model year 2011. However, NHTSA is not required to work with EPA to the extent it has on this joint rule. The collaboration of these two federal agencies came at the direction of the current administration to provide regulatory certainty and ensure that a clear set of rules was established for all automobile manufacturers.

NHTSA Improved the Analysis It Uses to Help Set CAFE Standards, and although Experts Still Expressed Some Concerns, They Lacked Consensus on Additional Improvements

NHTSA Evaluates Potential CAFE Standards Using the Volpe Model, Which Attempts to Simulate How Manufacturers Will Meet the Standards and Then Measures the Effects of the Standards

In part because NHTSA has previous experience in setting CAFE standards, we were asked to review any improvements NHTSA made to its process for setting CAFE standards.[44] We did so by looking in depth at NHTSA's regulatory impact analysis using the Volpe model, which has been used in previous rulemakings as well as the current proposed rule. It has been criticized by some experts in previous rulemakings for, among other things, a lack of transparency that limited public review. Because EPA is setting GHG emissions standards for the first time, we did not conduct a similar review of their modeling efforts using the OMEGA model.

The first key component of the Volpe model is a simulation of how manufacturers might comply with proposed CAFE standards. The "compliance simulation" of the Volpe model attempts to simulate each manufacturer's most cost-effective strategy to make its fleet comply with a more stringent CAFE standard by incorporating technologies until the manufacturer achieves compliance, exhausts all available technologies, or pays fines for noncompliance when it becomes more cost-effective than incorporating additional technologies. It relies on several key sources of data, including

- the "baseline vehicle fleet," a forecast of the vehicle models manufacturers will produce for sale in the U.S. in future model years;
- a list of available fuel-saving technologies, categorized into five groups;[45]
- estimates of the costs, effectiveness in reducing fuel consumption, applicability, and availability of these technologies; and
- pathways that estimate available fuel-saving technologies and the order in which manufacturers could take advantage of these technologies to most cost-effectively meet new CAFE standards.

This technology simulation is run for each vehicle model in the baseline fleet and produces an estimate of each vehicle's new fuel economy, weight, and total cost after the manufacturer has modified the vehicle in response to the CAFE standard. The compliance simulation's output is a forecast of model years 2012 through 2016 vehicles—namely, a re- engineered fleet of vehicles with new prices, fuel types, fuel-economy values, and weights to reflect the changes manufacturers would make to their vehicles to meet the proposed model year 2012 through 2016 CAFE standards. The data for each vehicle in the forecasted model year 2012 through 2016 fleet is then used in the second portion of the analysis.

This "calculation of effects" is the second key component of the Volpe model, which uses the compliance simulation data to estimate the costs and benefits of potential changes to the CAFE standard to manufacturers, consumers, and society as a whole. It uses a variety of data inputs, including fuel prices projected for the lifetimes of the vehicles in the fleet, the economic costs of fuel consumption, and damage costs for criteria pollutants.[46] This analysis produces information on the estimated benefits and costs of higher

CAFE standards, such as the benefit to consumers of fuel savings from driving more fuel-efficient vehicles, increases in new vehicle prices, changes in the number of vehicle miles traveled, and the societal benefits of reductions in carbon dioxide emissions. The estimated costs and benefits are used by NHTSA to set CAFE standards at a level that appropriately balances their costs and the benefits.

NHTSA Made Several Improvements to Its Analysis That Could Help NHTSA Better Estimate the Costs and Benefits of Increasing CAFE Standards

Transparency

To increase the transparency of inputs to the Volpe model for the 2012 through 2016 rulemaking, NHTSA used publicly available data to develop the model's baseline vehicle fleet. In previous rulemakings, NHTSA developed its baseline fleet by using confidential product plans submitted by manufacturers that described the vehicles manufacturers planned to sell in the U.S. in future years.[47] However, manufacturers submitted these plans to NHTSA as confidential business information, and NHTSA could not make these plans available to the public.[48] Comments submitted as part of prior CAFE rulemakings, as well as several experts we spoke to, indicated that the lack of transparency regarding NHTSA's use of product plans was troublesome because researchers could not replicate NHTSA's analysis. In developing their respective models for the joint rulemaking, NHTSA and EPA used a baseline fleet that drew primarily from public and commercially available information to make their analyses more transparent and provide additional validation of the results of their analyses.[49] Specifically, NHTSA and EPA relied almost entirely on information sources such as model year 2008 vehicle sales data, EPA's emission certification and fuel-economy database, and vehicle sales forecasts from several public sources.

There are several advantages of using public and commercially available data more extensively than product plans. First, federal regulatory analysis from OMB recommends that analyses be transparent to allow third parties to determine how the model produces its estimates and conclusions. By increasing the transparency of the baseline vehicle fleet, NHTSA allowed outside experts the opportunity to review the model's inputs and outputs and replicate the results of the model to better ensure that its analysis is thorough and sound. Second, because the submission of product plans is strictly voluntary, NHTSA has not consistently received complete information from all manufacturers with U.S. sales, which has inhibited its ability to forecast the future vehicle fleet across manufacturers using that data. Although several companies submit nearly complete product plans, others submit only partial plans, while still others do not submit any information. NHTSA also indicated it could save staff time by not having to correct errors in the manufacturers' submissions that NHTSA does receive.[50] Third, by using actual fuel-economy test data from model year 2008 vehicles, NHTSA would be able to use this verified fuel-economy information, rather than the estimates of the fuel-economy performance from vehicles' manufacturers.

Despite these advantages, there are some disadvantages to using the publicly available model year 2008 data to establish the baseline vehicle fleet. For example, by forecasting the model year 2012 through 2016 vehicle fleet using model year 2008 vehicle data, NHTSA and EPA's baseline includes vehicles that have been eliminated or for which production has been reduced, such as the Chrysler PT Cruiser and Hummer H2. It also does not include several vehicle models and technologies that manufacturers have recently introduced or plan to introduce, such as Ford's EcoBoost system (a package of engine technologies that in combination significantly improve fuel economy), the Honda Insight (a conventional hybrid), Chevrolet Volt (a plug-in hybrid electric vehicle), or Nissan's all-electric LEAF.[51] In addition to specific vehicles, NHTSA's baseline vehicle fleet forecast does not account for broad-scale changes to vehicle lines that manufacturers have started, such as Chrysler's plans to use Fiat power trains to offer small and medium-sized cars. Finally, NHTSA has found it difficult to determine, from either public or commercial sources, a number of specific data used in the baseline, such as information on electric power steering and reduced rolling-resistance tires. Consequently, NHTSA has had to use a small amount of data from product plans submitted in spring 2009 to fill these data gaps. NHTSA is also consulting with manufacturers regarding the possible release of model year 2010 or model year 2011 product plans that NHTSA could use in its development and analysis of the final model year 2012 through 2016 standards. Despite these disadvantages, NHTSA, EPA, and several experts we spoke to believe that the new transparency of its analysis outweighs the limitations of using public and commercially available data to establish its baseline.[52]

Technology Data and Economic Inputs

In the proposed model year 2012 through 2016 rule, NHTSA updated values for several data inputs in the Volpe model compared to its previous rulemakings, based on its own reviews of published research and several studies EPA conducted:

- *Technology data.* NHTSA reviewed the technology cost information used for model year 2011 CAFE standards, revising the cost estimates for several key fuel-saving technologies and reviewing and incorporating estimates of the effectiveness (i.e., fuel- saving improvements in mpgs) of these technologies (see figure 5 for an example of technology cost and effectiveness estimates). To determine technology cost estimates for the proposed rule, NHTSA and EPA reviewed the cost information in NHTSA's model year 2011 final rule, EPA's 2008 Staff Technical Report, and other sources.[53] The agencies revised component costs for several key technologies. For example, NHTSA revised the cost of turbocharging and downsizing an engine[54]—a cost range of $512 to $1,098, depending on engine type, compared to the range of $822 to $1,129 used for the model year 2011 CAFE standards—using data available from EPA's ongoing teardown study[55] with FEV, an automotive research, design, and development company. It also revised the costs of several other key technologies such as cylinder deactivation[56]—a cost range of $28 to $190, compared to the range of $306 to $400 used for the model year 2011 CAFE standards. However, despite this concerted effort, NHTSA and EPA were not able to make further refinements because the anticipated NAS study of vehicle technology was not completed on schedule.

- *Indirect costs to manufacturers.* NHTSA adopted research that EPA had contracted for to refine estimates of the indirect costs to manufacturers of manufacturing more fuel-efficient vehicles. These costs include research and development and marketing costs associated with the introduction of a new technology and give decision makers a more comprehensive view of the total costs a manufacturer would incur for implementing new technology than direct costs alone can provide. EPA supplemented an initial contractor report on this subject with an additional in-house study, which involved significant staff resources.

- *The social cost of carbon dioxide emissions.* NHTSA adopted an estimate of the damage resulting from carbon dioxide emissions that is more in-line with recent scientific and economic research, leading to a better reflection of the estimated benefits of increased CAFE standards related to reductions in GHG emissions. In the model year 2008 through 2011 light truck rule, NHTSA declined to include an economic value for reducing GHG emissions, citing the wide variation in published estimates of GHG emissions costs. However, a November 2007 federal court decision found that NHTSA's decision to not provide a monetized estimate of the benefit of reducing GHG emissions was arbitrary and capricious.[57] For the proposed model year 2012 through 2016 standards, NHTSA is using estimates of $5, $10, $20, $34, and $56 per metric ton of carbon dioxide— with an emphasis on the $20 value. These values, also adopted by EPA in its analysis, reflect the current administration's interim set of estimates of the social cost of carbon for agencies to use in regulatory analyses until a federal interagency working group develops a more comprehensive estimate for use in future economic and regulatory analyses.

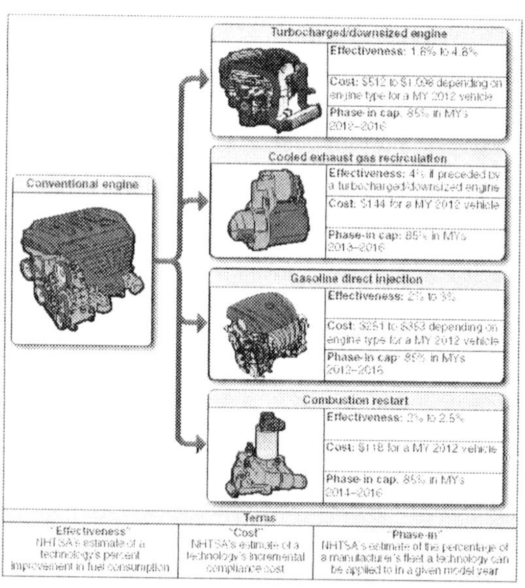

Source: GAO.

Figure 5. Example of Incremental Cost and Effectiveness Estimates for Technology Applications

- *Projected fuel prices.* NHTSA used the most recent and updated projections of fuel prices provided by the Energy Information Administration (EIA) to place a value on the fuel-saving costs and benefits of different CAFE standards. Among other things, the monetized benefits of the new CAFE standards are more sensitive to changes in fuel prices, meaning that the estimated benefits of more stringent CAFE standards will increase or decrease to a greater extent in response to changes in the price of fuel compared to changes in other variables. For the current proposal, NHTSA is using a range of prices from $2.50 in 2011 to $3.82 in 2030, which is consistent with the EIA's 2009 main fuel price projections,[58] and is focusing on an average retail gas price of $3.77 per gallon in 2007 dollars. In addition, NHTSA is reviewing the EIA's high and low fuel price projections to determine a range of potential costs and benefits, a best practice recommended by OMB guidance.[59] In projecting fuel prices, EIA considers recent and likely future developments in the world oil market, the effect of the current geopolitical situation on oil supply and prices, and conditions in the domestic fuel supply industry that affect pump prices. However, EIA projections have at times underestimated gas prices, most recently in 2008 during the price spike. Several experts we spoke to noted that gas prices are extremely difficult to predict.[60] However, most of the experts we spoke to also indicated that despite its limitations, EIA is the most credible source for projected fuel prices. Although EIA officials told us they do not issue guidance to agencies on how to use EIA projections in regulatory impact analyses, they expect agencies to consider that events EIA cannot predict will impact energy demand and fuel prices.

By applying the best research available, NHTSA should obtain better estimates of the benefits and costs of higher CAFE standards and allow standards to be set at a level better reflecting those benefits and costs.

More thorough Analysis

In line with OMB guidance on federal regulatory analysis, NHTSA conducted more thorough analyses in the proposed model year 2012 through 2016 standards than in previous CAFE rulemakings, including the model year 2008 through 2011 light truck rule. First, NHTSA tested and compared the benefits and costs of a greater number of CAFE levels set at different stringencies (also known as alternative scenarios) than it has in the past. By doing so, NHTSA gives decision makers a better picture of which level of CAFE standards provides the best balance between costs and benefits. NHTSA doubled the number of alternative CAFE scenarios it has tested from four to eight since the model year 2008 through 2011 light truck final rule. Specifically, NHTSA considered scenarios in which fuel- economy levels are increased at an annual average rate ranging from 3 to 7 percent, as well as scenarios in which the benefits are modified—for example, selecting a level at which the total costs of new CAFE standards are equal to their total benefits or a level that maximizes the net benefits of new CAFE standards to society. As a result, NHTSA was able to provide more comprehensive information for decision makers and increase public understanding of NHTSA's process for setting standards.

However, NHTSA also considered factors external to the model in determining the level of the proposed model year 2012 through 2016 standards. Although OMB guidance on regulatory analysis specifies that agencies should select the scenario that maximizes the net benefits of the regulatory action to society, NHTSA did not propose to select the "maximum net benefits" scenario as its preferred alternative for the standards in the proposed rule. Instead, NHTSA proposed to select a scenario in which CAFE standards increase at an average rate of 4.3 percent per year. According to NHTSA officials, that decision was justified because the four statutory factors that they must weigh when setting CAFE standards outweigh OMB guidance.[61] Several experts we spoke to said that NHTSA's decision was justified because selecting the "maximum net benefits" scenario would have resulted in CAFE standards that automobile manufacturers could not realistically meet without making significant tradeoffs. For instance, one expert thought manufacturers would have to change their fleet mix to build and sell smaller vehicles and would have to pass on substantial costs to consumers, which could reduce vehicle sales. In addition, another expert thought that if lead time is not sufficient, manufacturers will not be able to hire staff quickly enough to handle the additional work.

Additionally, as provided for in OMB guidance, NHTSA expanded its use of two types of uncertainty analysis, which differs from previous rulemakings. Specifically, relative to previous rulemakings, NHTSA expanded its sensitivity testing and probabilistic uncertainty analyses, both of which assess the uncertainty associated with key assumptions and inputs in its analysis, in comparison to previous rulemakings. NHTSA's sensitivity analysis and probabilistic uncertainty analysis test whether variability in the values of key model inputs would dramatically affect the costs and benefits of a potential

CAFE level. The variability of key inputs may arise from different estimates of credible studies or simply be the result of limited current knowledge. These sensitivity and uncertainty analyses provide decision makers with a sense of which potential CAFE level, despite the variability of key inputs, will best balance benefits and costs. In comparison to the model year 2008 through 2011 light truck rule, NHTSA's current sensitivity and probabilistic uncertainty analyses considered more case scenarios focusing on a number of critical inputs, including projections of fuel prices, the rebound effect, the value of reducing carbon dioxide emissions, and the military security benefits of reducing fuel consumption, of which variability in one input or a combination of inputs may affect the results of the overall analysis.

Experts Continue to Have Concerns about the Model, but Do Not Agree on How to Further Improve Inputs to the Model

As part of this work, we spoke with a number of experts familiar with the Volpe model about their assessment of the data used in the model. Although they provided criticism, they did not agree on what needed to be improved (see app. I for information on experts with whom we consulted). In general, nearly all of the experts we spoke to offered some critique of the model and its data. For instance, some, but not all, experts said that NHTSA was too cautious in updating the values for variables such as the social cost of carbon dioxide emissions, given the state of current research. These experts said that NHTSA was underestimating the social cost of carbon dioxide emissions, which would lead to an underestimation of the benefits of CAFE standards and the establishment of standards set at a lower than ideal level. However, we could not find general consensus among experts we spoke to that NHTSA should have modified values for specific variables or made other improvements to the model. For example, NHTSA used a lower value for the rebound effect[62] (10 percent) to more closely align with values identified in recent research. Several experts thought that NHTSA should have adopted the value (5 percent) identified in the research, which was even lower than what NHTSA used, while others thought that NHTSA's more cautious approach was appropriate until additional studies using different data sets verified the findings.

We did find considerable controversy among experts over the potential safety impact of weight reduction in vehicles—much more so than for other variables assessed in the Volpe model.[63] While some experts stated that manufacturers could safely reduce vehicle weight while maintaining the size of the vehicle by substituting lightweight but durable materials for heavier materials (material substitution), other experts maintained that any effort to reduce vehicle weight would adversely affect safety. Two studies, one developed by NHTSA (Kahane study)[64] and a second conducted by an automotive engineering consulting firm (Dynamic Research, Inc., study),[65] came to different conclusions on this issue, and to date, no subsequent study has been conducted in a manner designed to resolve the conflict. DOE has sponsored research through the Lawrence Berkeley National Laboratory that examines the relationship between vehicle weight and driver casualty risk using police-reported crash data and CAFE compliance records, but given the high level of ongoing controversy, this approach may not satisfy all the experts invested in this issue. In addition, neither the Kahane study nor the Dynamic Research, Inc., study were able to assess directly how material substitution as a particular approach to weight reduction could affect safety because the vehicles analyzed in the two studies were limited to model years 1985 through 1999. During this period, CAFE standards were not attribute- based, and manufacturers had a greater incentive to improve fuel economy by reducing vehicle size rather than by reducing vehicle weight through material substitution. In addition, several experts noted that by using the Kahane study in its current work, NHTSA may be overestimating the safety implications of higher CAFE standards because the study does not consider technology solutions like material substitution as an option that could improve fuel economy without negatively affecting safety. Because NHTSA accounts for the safety effects of proposed standards by estimating their safety implications, relying on this research in the future could result in standards being set at a lower level. In the past, concerns about safety have prevented non-attribute-based CAFE standards from being increased.

We also learned from experts that vehicle safety is challenging to address because the safety tradeoff between larger, heavier vehicles and smaller, lighter vehicles does not lend itself to a clear policy solution. Generally, larger and heavier vehicles, which enhance the safety of their passengers as a result of their size and weight, pose a greater safety threat to other vehicles on the roadways than smaller, lighter cars do. Conversely, although smaller, lighter cars pose less of a threat to other vehicles on the road, they cannot provide the same degree of safety to their passengers that larger, heavier vehicles do. The degree of difference in the size and weight of vehicles has some bearing on passenger safety: larger, heavier vehicles provide their passengers safety benefits and impose on others safety costs, while smaller, lighter vehicles provide others safety benefits and impose on their passengers safety costs.

Several experts with whom we spoke thought that additional research was needed to better understand the relationship between vehicle size, weight, and safety, as well as to identify how best to reduce the weight of vehicles in a manner that creates the least risk. Experts recommended several different methodological approaches to assess this relationship, including future studies that examined material substitution in accident outcomes once vehicles with this technology became more prevalent in the fleet. Others recommended the use of computer crash simulation modeling to identify best practices in the use of material substitution.

Largely Due to Resource Constraints, NHTSA Has Limited Plans to Assess the Model Year 2008 through 2011 Light Truck CAFÉ Standards or Key Data Used to Develop the Standards

Federal agencies can use retrospective analyses of rulemakings to help determine the extent to which the expected costs, benefits, and goals of a regulation are being realized.[66] A retrospective analysis of CAFE standards could help NHTSA and Congress determine the extent to which goals of the standards—such as improvements in fuel economy—are being met and provide insight into ways to improve the standards. In addition, a retrospective analysis of key data inputs could help determine if there are systematic issues with the estimation of those data and identify means to improve the data in the future. EPA officials noted that they have used retrospective analyses of other regulatory programs to assess the accuracy of program costs. For example, in 2002, EPA issued a retrospective cost analysis of a large number of light-duty vehicle criteria pollutant standards and mobile source fuel standards implemented between 1992 and 2001.[67] However, because EPA has not previously issued GHG emissions standards for automobiles, it would not be able to conduct these types of analyses for GHG emissions standards at this time.

With respect to the model year 2008 through 2011 light truck CAFE standards, the following retrospective analyses could be conducted by NHTSA:

- An overall analysis of the standards to determine the extent to which the new, footprint-based standards met intended goals (e.g., increases in fuel economy and reductions in fuel consumption). As the proposed model year 2012 through 2016 CAFE standards are also to be based on vehicle footprint, this analysis could help determine if the move to the footprint based standard provided the intended benefits or imposed unexpected costs. [68]

- An analysis of the accuracy of key data inputs, including the baseline fleet and technology cost estimates. NHTSA has been criticized in the past for not adequately estimating these two sets of data, which provide crucial information for determining the effects of the proposed standards, and thus need to be as accurate as possible.
  Although NHTSA officials we spoke with recognize the value of these analyses and hope to conduct them, they report that resource limitations have prevented them from doing so in the past and will prevent them from doing so in the near future. In addition, NHTSA is not required to do any of these analyses. A discussion of NHTSA officials' responses regarding retrospective analyses and the resource limitations that have prevented them from being conducted follows:

- *Model year 2008 through 2011 light truck standards*. NHTSA staff said that such retrospective analysis of the model year 2008 through 2011 light truck standards would be worthwhile and informative. However, according to NHTSA officials, in recent months the agency has devoted all of its dedicated CAFE staff's time to the proposed model year 2012 through 2016 CAFE rule and, as a result, has not been able to devote resources to conducting a retrospective analysis.[69] In addition, given that NHTSA staff said that the agency is being asked by a majority of commenters addressing the subject to begin working on CAFE standards beyond model year 2016 as soon as possible, they may not be able to work on a retrospective analysis once the model year 2012 through 2016 standards are finalized and released. However, a number of experts we interviewed said that NHTSA should conduct such an analysis in order to provide insight into the standards and their actual effects.

- *Manufacturers' sales data*. While NHTSA told us that it would like to look back at manufacturers' actual sales as a means to assess the accuracy of the product plans that manufacturers submitted and that NHTSA used as the baseline fleet in setting model year 2008 through

2011 light truck standards, it said that it has no definitive plans for conducting this analysis in the near future. NHTSA officials cited a lack of resources in the agency for not conducting such an analysis. In addition, because 2008 sales were an anomaly—they were unusually low given the economic downturn— officials thought a study of the extent to which actual 2008 sales were in- line with the forecasted sales for 2008 that were used to set those standards would be of little value. However, an analysis of actual future years' sales against the estimated sales of the baseline fleet used in the rulemaking would be of value, as it would help validate data and potentially identify means to improve fleet forecasts in future CAFE rulemakings.

- *Cost estimates of technology*. NHTSA officials also told us that an assessment of the cost estimates of technology used in previous analyses would be valuable. However, NHTSA staff also said that such an analysis would be challenging, as it is hard to get accurate data on the actual cost of technology components. This is because these components are either sold directly to, or produced by, automobile manufacturers, meaning that there is no clear, public historical data on their sales price. However, while some experts with whom we spoke recognized the challenges in conducting such an analysis, they thought that such an assessment would provide value and recommended several different approaches for conducting this type of analysis. For example, some experts suggested that costs could be validated through a vehicle teardown program, such as the type of project EPA initiated last year, or through an analysis of sales data and technology that manufacturers incorporated into recent models to comply with increased standards. While these studies could potentially impose large resource demands, they would also potentially help improve the cost of technology assumptions in future CAFE rules, helping to create standards that more accurately reflect costs and benefits.

CONCLUSIONS

Because CAFE and GHG emissions standards are closely related and automobile manufacturers will be subject to both, close collaboration between

NHTSA and EPA can minimize compliance costs to the industry and ensure harmonized standards. Furthermore, regardless of how the government may set any future standards—jointly or independently—a continued partnership between the two agencies can help assure fiscal responsibility by leveraging—rather than duplicating—federal efforts and resources, including expertise and human capital costs. However, the current level of collaboration between NHTSA and EPA, which stems from the joint rulemaking process the agencies undertook at the discretion of the current administration, is not set in law or otherwise required. If NHTSA and EPA do not collaborate closely on future standards, there is a risk that the standards may not be harmonized, which would lead to increased compliance costs for manufacturers; the standards may not reflect the expertise of both agencies, such as the vehicle power train technology and environmental expertise of EPA and vehicle safety expertise of NHTSA; and the goals that the standards are attempting to accomplish may not be met. Also, the standards may not accurately reflect the best estimates of key costs and benefits, thus imposing added costs on the economy or failing to provide as large benefits to society as the standards could.

In addition, this is the first joint rulemaking conducted between these agencies, and NHTSA and EPA are under tight time frames to set the standards. However, the agencies are not documenting the processes being used. If NHTSA and EPA must collaborate on future standards, staff may spend additional time recreating these processes—ones which appear to be working effectively—and relearning how best to interface with one another's leadership structure, management processes, and research activities. As a result, the two agencies may not share their respective expertise and resources as well, potentially leading to inefficiencies, less thorough and rigorous regulatory analyses, and standards that may not be effectively harmonized or developed with similar time frames.

NHTSA has not yet conducted—nor does it have plans to conduct—a full and formal analysis of the effectiveness and outcomes of its adoption of the footprint-based CAFE standards for light trucks. Also, it has no plans to assess the accuracy of key data inputs used to set these standards, even though it is now proposing a footprint-based approach for passenger vehicles as well. Conducting these types of analyses can help policymakers determine whether anticipated benefits and costs have been realized and identify corrections in or improvements to existing programs. NHTSA is not required to conduct such analyses and has limited staff and resources to devote to this effort. As a result, it is not clear if the new standards have met goals that NHTSA intended—such as fuel savings and improved safety outcomes—and if the move to the footprint-based standards was worthwhile. Furthermore, NHTSA does not know how well it estimated key data inputs that help determine the level at which standards are set, including technology costs; whether manufacturers used the types of technologies NHTSA expected in order to comply with new standards; and whether baseline fleets matched the vehicle mix actually sold. Consequently, agency officials cannot learn from the past and make adjustments to the process, such as seeking different data sources, to ensure that future standards are based on the most accurate data available.

Given the importance of safety in setting CAFE standards, ensuring that decision makers and the public have the most accurate information on the relationship between vehicle size, weight, and safety will be important if the standards are to be changed in the future. In addition, the data inputs that NHTSA and EPA use to help set and analyze the effects of the proposed model year 2012 through 2016 standards should be based upon the best available research and reflect a consensus among experts and stakeholders. Given the controversy among experts and the increasing availability of material substitution—an advancement in technology to reduce weight that could compensate for safety effects—new research could help to answer questions regarding the extent to which weight can be reduced without affecting safety and whether there are best practices for employing material substitution.

Finally, while other sources of technology costs were used in developing CAFE and GHG emissions standards, the 2002 NAS work on technology costs was generally viewed by a wide range of stakeholders and experts as being thorough and unbiased. Congress authorized NHTSA to contract with NAS at 5-year intervals until 2025 so that the agency would have current information available to set future standards. However, if NHTSA cannot ensure that this work is available in time to support analysis in future rulemaking, this study, and the federal money that sponsored it, will be wasted.

Recommendations for Executive Action

Based on our review, we are making five recommendations. We recommend the following to NHTSA and EPA:

- NHTSA and EPA should document the process used in this joint rulemaking to establish a roadmap for any future rulemaking efforts and facilitate future collaboration. In addition, NHTSA and EPA should publish this documentation in order to increase transparency.

- To ensure continued collaboration and an enhanced relationship in any future CAFE and GHG emissions rulemakings, NHTSA and EPA should enter into a Memorandum of Understanding with one another in which the agencies agree to continue their enhanced partnership in any future CAFE and GHG rulemakings.
- NHTSA and EPA, with input from key stakeholders, should conduct or sponsor new research on safety and its relationship to vehicle size and weight, given the controversy and lack of consensus regarding the relationship between vehicle size, weight, and safety and the emergence of new strong-but-lightweight materials among experts and stakeholders.

In addition, we are recommending the following to NHTSA:

- NHTSA should conduct and document a retrospective analysis of the model year 2008 through 2011 light truck standards, given the potential impact of CAFE standards on the automobile industry and consumers. In addition, we recommend that NHTSA identify opportunities to evaluate the accuracy of key estimates, such as technology costs, used to determine the model year 2008 through 2011 light truck standards. As EPA has experience conducting retrospective analyses of regulatory programs, NHTSA should consider involving EPA in this process.
- NHTSA should set delivery time frames for future NAS studies to ensure the availability of these studies in a time frame useful for incorporation in NHTSA's regulatory analyses.

Agency Comments and Our Evaluation

We provided a draft copy of this chapter to the Department of Transportation and the Environmental Protection Agency for their review. We also provided a relevant section of the report to the Energy Information Administration, and officials confirmed that information characterizing EIA's fuel price projections was accurate.

EPA provided a written response, which is reproduced in appendix III. In its response, EPA agreed with our characterization of NHTSA and EPA's collaboration on setting CAFE and GHG emissions standards and with our recommendations. In addition, EPA provided technical comments via email which we incorporated as appropriate.

DOT provided its response by e-mail and generally agreed with the report's recommendations. NHTSA also provided technical comments, and while we incorporated a number of these comments, others offer an opportunity for additional discussion. First, NHTSA suggested that our first two recommendations—(1) that NHTSA and EPA document the process used in this joint rulemaking, and (2) that NHTSA and EPA sign a Memorandum of Understanding to continue this enhanced partnership—apply only if future rulemakings are conducted jointly. We did not make this change. Given NHTSA and EPA's successful collaboration on CAFE and GHG emissions standards, we believe continued collaboration will help ensure that federal resources and expertise are leveraged efficiently and effectively—regardless of whether future administrations continue to issue both sets of standards jointly, separately, or pursue only CAFE or GHG emissions standards.

Second, in our discussion of the impact of the appropriations ban from fiscal years 1996 through 2001 that prevented NHTSA from conducting work on CAFE issues, we noted that NTHSA lost staff with relevant expertise and did not begin to hire additional automotive engineers until summer 2009. We looked into this issue because in our 2007 report. NHTSA officials told us they needed additional staff with expertise in automotive engineering and computer modeling to assist in developing technology cost and effectiveness estimates, as well as other tasks, to prepare for future changes in CAFE standards. NHTSA commented in response to this draft that the prohibition did not prevent DOT from sustaining relevant engineering, energy, and environmental expertise, and that after 2001, NHTSA leveraged DOT's expertise. NHTSA also commented that in our current review, we did not examine broader staff capabilities within DOT. We agree that this information is important. However, we were not able to confirm the extent to which NHTSA leveraged DOT's expertise because NHTSA did not provide this information. We continue to believe that NHTSA and EPA have different expertise and resources—ones that likely cannot be replicated efficiently at both agencies but that are crucial for the development of balanced, effective standards for cars and light trucks, and therefore we did not revise the report.

Appendix I: Objectives, Scope, and Methodology

To describe the proposed corporate average fuel economy (CAFE) and greenhouse gas (GHG) emissions standards, we analyzed documentation related to the rulemaking, such as the May 2009 Notice of Upcoming Joint Rulemaking, September 2009 Notice of Proposed Rulemaking, and associated preliminary regulatory impact analyses from both agencies. We analyzed these documents to summarize the structure of each set of standards, describing how the National Highway Traffic Safety Administration (NHTSA) and the Environmental Protection Agency (EPA) harmonized the standards and areas in which there are differences between the standards, such as certain types of flexibilities like temporary lead-time mechanisms. We also summarized related legislation that establishes CAFE fines and summarized EPA's authority under the Clean Air Act to assess fines for noncompliance with GHG standards, to describe the penalties that NHTSA and EPA will apply for noncompliance with the new standards.

To describe NHTSA's and EPA's processes for setting proposed model year 2012 through 2016 CAFE and GHG emissions standards, we reviewed and analyzed relevant rulemaking documents, such as the Notice of Proposed Rulemaking and the legislation establishing CAFE standards and EPA's authority to regulate GHG emissions, noting the types of analyses each agency was allowed to conduct under its individual legal authority. We analyzed documentation related to the analyses the agencies conducted. We also interviewed agency officials and reviewed documentation from NHTSA and EPA related to the work they conducted in setting the standards. To describe how the agencies collaborated with one another to issue the standards, we analyzed these interviews and documentation against GAO criteria for evaluating communication and coordination among federal agencies. Through interviews with officials and by reviewing research each agency developed as part of the rulemaking, we identified the expertise and resources each agency brought to bear in the development of the standards.

To evaluate the improvements made to NHTSA's regulatory impact analyses used in setting CAFE standards, we reviewed relevant documentation, including NHTSA's Preliminary Regulatory Impact Analysis on model year 2011 CAFE standards for passenger cars and light trucks and for the proposed model year 2012 through 2016 standards. We also conducted literature searches for research on fuel economy published since 2007—the year of our last report on CAFE standards. We interviewed NHTSA officials and staff at the Volpe National Transportation Systems Center, as well as automobile industry stakeholders—including domestic and international automobile manufacturers; an association representing original equipment suppliers; vehicle technology specialists at national laboratories and academic research centers; and independent experts on vehicle technology, transportation, and modeling. We identified these experts through several approaches:

- About half of the experts we contacted had assisted us in our 2007 review of CAFE standards. Several of these experts were members of the current or 2002 National Academy of Sciences (NAS) committee, while others had been recommended by members of the NAS committee or NHTSA.
- We conducted internet searches to identify experts publishing recent research on fuel economy, GHG emissions, economic modeling, and other issues.
- We asked experts participating in our work for recommendations.

We also pursued a more in-depth analysis from stakeholders about safety and vehicle weight by reviewing the methodology of several key studies and interviewing engineers and other organizations with specific expertise in safety and vehicle design, such as the Insurance Institute for Highway Safety and experts from National Laboratories. We also interviewed officials from the Energy Information Administration (EIA) to review gasoline price projections that are used in the Volpe model. To evaluate NHTSA's processes for obtaining and validating data on automobile manufacturer product plans and cost data on fuel-saving technologies, we analyzed NHTSA documentation against GAO criteria for developing, managing, and evaluating cost estimates and for assessing data reliability. To evaluate NHTSA's processes for estimating the costs and benefits of improved vehicle fuel economy in the Volpe model, we analyzed NHTSA documentation against federal guidance for conducting regulatory and economic analyses and GAO guidance for conducting benefit-cost analyses.

To determine the steps NHTSA has taken to analyze the effects of the model year 2008 through 2011 light truck standards, we reviewed and analyzed the Energy Independence and Security Act, NHTSA's final rulemaking on the model year 2008 through 2011 CAFE standards for light trucks, and the data used to set these standards. We interviewed NHTSA officials to determine whether NHTSA has conducted analyses to assess the outcomes of these standards—for example, improvements in vehicle fuel economy and gallons of oil saved—and requested documentation of any analyses. To determine the steps NHTSA has taken to assess the accuracy of input data and assumptions used in developing the model year 2008 through 2011 CAFE standards—particularly assumptions related to cost estimates of technology and manufacturer product plans—we interviewed NHTSA officials and requested documentation of any analyses as appropriate. For example, we assessed whether NHTSA compared data that estimated the costs of fuel-saving technology to actual cost data from 2008. We also interviewed outside experts on options NHTSA could use to conduct such an analysis and the benefits and tradeoffs of doing so. Finally, we reviewed and analyzed these interviews and documentation against GAO guidance for program evaluation.

We conducted this performance audit from June 2009 to February 2010, in accordance with generally accepted government auditing standards. Those standards require that we plan and perform the audit to obtain sufficient, appropriate evidence to provide a reasonable basis for our findings and conclusions based on our audit objectives. We believe that the evidence obtained provides a reasonable basis for our findings and conclusions based on our audit objectives.

Appendix II. The CAFE Compliance and Effects Modeling System

As part of its regulatory impact analysis of potential CAFE standards, NHTSA uses the CAFE Compliance and Effects Modeling System (commonly known as the Volpe model) developed by the Volpe National Transportation Systems Center to estimate the following: (1) the most cost-effective strategy for automobile manufacturers to respond to proposed CAFE standards and (2) the impacts, such as reduced fuel consumption, increased vehicle prices, and reduced emissions, proposed CAFE standards will have on consumers, manufacturers, and society. For a visual description of the Volpe model's analysis, see figure 6.

Input Files Used in the Volpe Model

The Volpe model's analysis relies on a number of data inputs, including, among other things, a list of the automobile manufacturers producing vehicles for sale in the U.S. during the period covered by a CAFE rulemaking, a list of fuel-saving technologies[70] and their estimated cost and effectiveness in reducing fuel consumption, simulated alternative CAFE scenarios (i.e., CAFE standards set at range of levels), economic inputs such as the estimated social cost of carbon dioxide emissions and the rebound effect (a phenomenon in which individuals drive more because improving a vehicle's fuel economy effectively lowers the cost per mile of operating that vehicle), and the emissions rates of various pollutants. These data are contained in several input files that are entered into the Volpe model.

The Compliance Simulation

The Volpe model's compliance simulation demonstrates how each automobile manufacturer could attempt to comply with a higher CAFE standard by adding fuel-saving technologies to its vehicle fleet until that level is achieved. Using the information provided in the scenario input file, the Volpe model applies fuel-saving technologies in order of cost- effectiveness and ease of implementation to the vehicle models forecasted in the baseline to simulate how a manufacturer could make progress toward compliance with new CAFE standards.

Definition of the Baseline Vehicle Fleet

The compliance simulation begins with a forecast of the U.S. vehicle fleet in future model years, which represents the baseline vehicles (including estimates of the volumes and prices of individual vehicle models) manufacturers could modify with fuel-saving technologies to comply with the model year 2012 through 2016 CAFE standards. For the model year 2012 through 2016 rulemaking, the baseline vehicle forecast was developed using public model year 2008 vehicle sales data, vehicle sales forecasts from EIA, forecasts of the relative sales of cars and trucks by manufacturer and market segment from CSM-Worldwide, EPA's emission certification and fuel-economy database, vehicle and technology information from Edmunds.com, Motortrend.com, and Ward's Automotive, and to a more limited extent than in previous rulemakings, confidential business plans provided by automobile manufacturers upon request.[71]

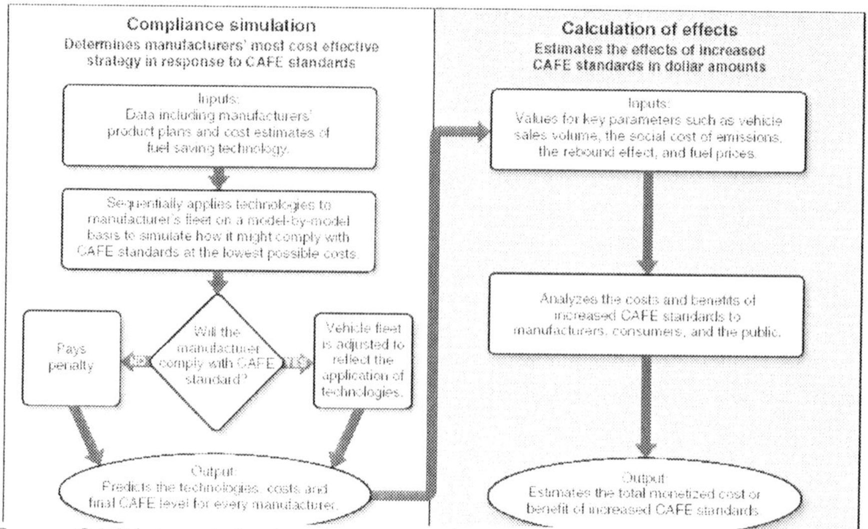

Source: GAO interpretation based on past use by NHTSA.

Figure 6. The Volpe Model

Application of Vehicle Technologies

Using the baseline vehicle fleet, the Volpe model then simulates how each manufacturer could apply fuel-saving technologies to each vehicle model in its fleet to comply with the model year 2012 through 2016 CAFE standards in the most cost-effective manner. Prior to this simulation, NHTSA estimated the cost, effectiveness in reducing fuel consumption, applicability in terms of vehicle subclass,[72] availability by model year, learning rate,[73] and fleet penetration[74] of each technology considered in the compliance simulation and included this information in the technology input file. Technologies are categorized within one of five technology groups,[75] and each technology group has a corresponding "decision tree" which displays the sequence in which NHTSA estimates a manufacturer would apply technologies to the vehicle models in its fleet. For example, a manufacturer could apply electrical power steering, improved electrical accessories, 12-volt micro-hybrid technology, a belt integrated starter generator, and a crank integrated starter generator to a subcompact car using the decision tree for the electrification/accessory technology group. Each technology is positioned along a decision tree according to its estimated incremental cost and fuel-economy improvement, taking into account technologies that have already been applied. Before applying new technologies to a vehicle, the Volpe model first carries over any technologies that were present during the previous model year. Then, proceeding along each technology group's decision tree, the Volpe model determines the applicability and availability of each technology to every vehicle model. If the phase-in limit for a particular technology has been reached and it is no longer available, the Volpe model proceeds to the estimated next-best technology.[76] See figure 7 for a visual description of the process by which the Volpe model determines the applicability and availability of a given technology.

The model repeats this process for each technology group, and then selects the technology with the lowest effective cost—that is, the technology that provides the greatest private benefits with the lowest cost. The compliance simulation continues to apply technologies to each manufacturer's fleet using this approach until (1) the manufacturer's fleet is estimated to be brought into compliance with the CAFE standard for a given model year, (2) the manufacturer has exhausted all the technology options for its fleet, or (3) the Volpe model estimates that it would be more cost-effective for the manufacturer to pay the associated CAFE fines than to apply additional technology to its fleet. The Volpe model accounts for multiyear planning, through which a manufacturer may apply more technology than necessary in earlier model years in order to carry those technologies forward into future model years and thereby avoid applying other more expensive technologies. When the Volpe model has brought each manufacturer's fleet to one of the three outcomes listed above, the compliance simulation loop ends.

Forecast of the New Vehicle Fleet

The compliance simulation produces an output file that shows, for each vehicle in a manufacturer's fleet, which technologies were included in a vehicle model before the simulation was run, which technologies were skipped in favor of other technologies, and which technologies had been applied to vehicles at the simulation's end. The output file also shows the changes in vehicle weight, improvement in fuel economy, and incurred cost resulting from the technologies applied during the compliance simulation, as well as the total cost of any civil penalties incurred by each manufacturer. At this point, the Volpe model has a new fleet of vehicles with new prices, fuel types (gasoline or diesel), fuel-economy values, and curb weights to reflect how NHTSA estimates manufacturers will apply fuel-saving technologies in response to the CAFE requirements.

The Calculation of Effects

Following the compliance simulation, the Volpe model's calculation of effects component estimates the impact of the fuel-economy improvements made to vehicles to meet new CAFE standards on energy consumption, greenhouse emissions, and other factors. Using the forecasted vehicle fleet (i.e., the output of the compliance simulation), the Volpe model estimates the lifetime travel, fuel consumption, and carbon dioxide and criteria pollutant emissions[77] resulting from the application of technologies to meet higher CAFE standards for each vehicle in the U.S. fleet over its anticipated life span. After calculating the effects for individual vehicle models, the Volpe model aggregates these effects for all the vehicles in a CAFE class produced during each model year affected by a proposed standard.

Costs, Benefits, and Effects of More Stringent CAFE Standards

The Volpe model measures the effects of increased CAFE standards by calculating the difference in the value of a variable (e.g., gallons of fuel consumed) under the baseline (model year 2011) CAFE standard and its value under a new CAFE standard. These effects include but are not limited to

- reductions in greenhouse gas emissions—increasing CAFE standards will reduce gasoline consumption and the amount of petroleum refined, which will reduce emissions of greenhouse gases;
- higher or lower emissions of air pollutants;
- potential increases in new vehicle prices;
- social value of fuel savings, which is the annual value of fuel savings over the entire expected lifetimes of vehicle models whose fuel economy is improved;
- economic benefits from reduced petroleum imports;
- valuing changes in environmental impacts (i.e., the Volpe model estimates changes in damage costs caused by carbon dioxide emissions); and
- social costs of added driving.

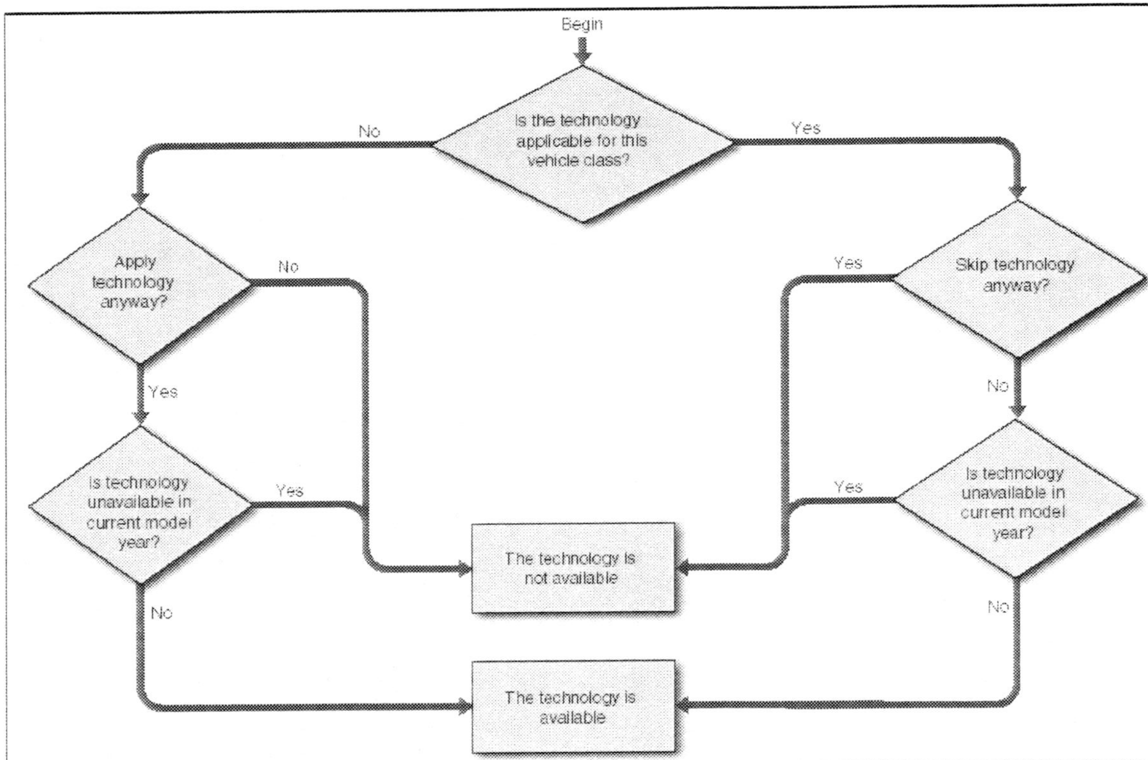

Source: GAO adaptation of the Volpe model's technology applicability determination.

Figure 7. The Volpe Model's Determination of Technology Applicability and Availability

Appendix III. Comments from the Environmental Protection Agency

UNITED STATES ENVIRONMENTAL PROTECTION AGENCY
WASHINGTON, D.C. 20460

FEB 16 2010

OFFICE OF
AIR AND RADIATION

Ms. Susan Fleming
Director, Physical Infrastructure
Government Accountability Office
441 G. St., N.W.
Washington, D.C. 20548

Dear Ms. Fleming:

Thank you for the opportunity for the Environmental Protection Agency (EPA) to respond to the Government Accountability Office's (GAO) draft report "*NHTSA and EPA's Partnership for Setting Fuel Economy and GHG Emission Standards Improved Analysis and Should be Maintained,*" including the recommendations included in the report.

EPA fully understands its responsibilities under the Clean Air Act to protect human health and welfare, including the responsibilities which flow from the April 2007 Supreme Court decision in *Mass. v. EPA* with respect to climate change and greenhouse gases (GHG). As discussed in the GAO report, one of the first formal actions undertaken by EPA in response to the Court's decision was the September 2009 Joint Notice of Proposed Rulemaking (NPRM) for light-duty vehicles from EPA and the National Highway Traffic Safety Administration (NHTSA). In that action, EPA proposed the federal government's first GHG standards for light-duty vehicles, and NHTSA proposed closely coordinated fuel economy standards for the same vehicles covering the same vehicle model years (2012 to 2016). This joint NPRM demonstrates the results of several months of close collaboration between EPA and NHTSA.

GAO's draft report includes a review of the collaboration between EPA and NHTSA over the past year as the two agencies worked together to develop coordinated programs which will reduce GHG emissions, improve vehicle fuel economy, provide the industry with coordinated regulatory programs which can be met by a single national vehicle fleet, result in very large oil savings and GHG reductions, and provide significant positive lifetime fuel savings which far exceed the projected increase in vehicle costs.

GAO presents four recommendations in the draft report which are directed at both EPA and NHTSA. EPA agrees with each of these recommendations, and below we respond to each recommendation.

GAO recommendation 1: EPA and NHTSA should document their collaborative rule development process.

Comment: EPA agrees with this recommendation. After the completion of our current rulemaking, we will work with our colleagues at NHTSA to document the process by which we worked together for both the NPRM and the final rule in an effort to learn from what we did right and look for opportunities for improvement in future collaborations.

GAO recommendation 2: EPA and NHTSA should formalize their collaborative relationship in order to ensure cooperation on future GHG and CAFE rule development.

Comment: EPA agrees with this recommendation. Given the close relationship between vehicle CO_2 emissions and vehicle fuel economy, it is important that EPA and NHTSA work closely together in the future to ensure that the two agencies' independent regulatory authority is closely coordinated to ensure that each agency's responsibilities are met in a way that ensures a coordinated regulatory regime for the automotive industry. EPA will work with our colleagues at NHTSA to identify ways to strengthen our current collaboration for future regulatory actions, including exploring the possibility of a formal Memorandum of Understanding as suggested by GAO, or some other form of formal agreement.

GAO recommendation 3: NHTSA and EPA, with input from key stakeholders, should conduct or sponsor research on safety and its relationship to vehicle size and weight.

Comment: EPA agrees with this recommendation. EPA has a responsibility under the Clean Air Act to consider the potential impacts of vehicle emission standards, including impacts on vehicle safety. The issue of vehicle size and mass and the potential impact of future GHG standards on vehicle safety is critically important, and EPA intends to work closely with our colleagues at NHTSA to improve the scientific understanding of these issues in the context of future improvements in vehicle GHG performance and fuel economy.

GAO recommendation 4: NHTSA should conduct a retrospective analysis of the 2008-2011 CAFE rule, and NHTSA should consider involving EPA in this analysis.

Comment: EPA has performed retrospective analysis of a number of our past mobile source rulemakings (e.g., fuel standards, light-duty vehicle standards, and heavy-duty engine standards), and we have found such analysis to be instructive. To the extent our colleagues at NHTSA decide to undertake a retrospective analysis of the 2008-2011 CAFE standards, EPA will assist in any way that would be constructive.

In addition to our comments discussed in this letter, EPA has also provided to GAO a number of suggested editorial changes to the draft report as a mark-up of the draft. Overall, EPA believes the draft report provides an accurate description of the collaboration between EPA and

NHTSA in the development of the proposed 2012-2016 model year light-duty vehicle GHG and CAFE standards.

Once again, thank you for the opportunity to review this draft report.

Sincerely,

Margo Tsirigotis Oge
Director
Office of Transportation and Air Quality

End Notes

[1] Pub. L. No. 94-163, 89 Stat. 871 (Dec. 22, 1975).

[2] EPCA requires EPA to measure fuel economy using procedures prescribed by EPA. Manufacturers are required to provide EPA with all data needed to determine their CAFE level for each model year. For the 2001 model year and later, EPA must generate a summary report of fuel economy values contained in the CAFE calculation. NHTSA uses the end-of-year report from EPA, along with any credits earned in the past or borrowed from the future, to determine if a manufacturer's fleet is in compliance for that model year, and if not, NHTSA is responsible for notifying the manufacturer of any required fine.

[3] See e.g. Pub. L. No. 104-50, §330, 109 Stat. 436, 457 (Nov. 15, 1995).

[4] Unreformed refers to the single CAFE standard that existed prior to the adoption of attribute-based standards.

[5] Pub. L. No. 110-140, 121 Stat 1492 (Dec. 19, 2007).

[6] DOT, *Average Fuel Economy Standards Passenger Cars Light Trucks Model Years 2011- 2015*, 73 Fed. Reg. 24352 (May 2, 2008).

[7] The previous administration did not publish a final rule for all five model years. For the final rule for model year 2011 see 74 Fed. Reg. 14196 (March 30, 2009).

[8] *Massachusetts v. Environmental Protection Agency*, 549 U.S. 497 (2007).

[9] Section 202 of the CAA requires EPA to regulate the emission of air pollutants from mobile sources which cause or contribute to air pollution which may reasonably be anticipated to endanger public health or welfare. The Endangerment Finding represents the EPA Administrator's conclusion that four greenhouse gases meet the threshold requirement for regulation under Section 202: carbon dioxide, methane, nitrous oxide, and hydrofluorocarbons. Endangerment and Cause or Contribute Findings for Greenhouse Gases under Section 202(a) of the Clean Air Act, 74 Fed. Reg. 66496 (Dec. 15, 2009).

[10] A joint resolution of disapproval under the Congressional Review Act has been introduced in the Senate. S.J. Res. 26, 111th Cong. (2010). If enacted, this resolution would void EPA's finding that GHGs endanger public health and welfare and cause or contribute to air pollution. Similarly, several bills have been introduced which would exclude GHGs from being defined as "air pollutants" subject to regulation under section 202 of the Clean Air Act. See H.R. 4396, 111th Cong. (2010); H.R. 4752, 111th Cong. (2010). In addition, numerous lawsuits have been filed challenging the endangerment finding.

[11] The CAA generally allows one set of federal standards for new motor vehicle emissions and preempts states from adopting or enforcing their own standards. However, it also requires the EPA Administrator to waive this preemption provision for any state that adopted certain emission standards for new motor vehicles prior to March 1966 if the state makes a finding

that its standards are as protective, in the aggregate, as applicable federal standards. California is the only state which has met the requirement for obtaining a waiver.

[12] Cent. Valley Chrysler-Jeep, Inc. v. Goldstene, 529 F. Supp. 2d 1151 (E.D. Cal. 2007); Green Mt. Chrysler Plymouth Dodge Jeep v. Crombie, 508 F. Supp. 2d 295 (D. Vt. 2007). In both cases, the courts held that that federal preemption did not apply to preclude the state regulations.

[13] California State Motor Vehicle Pollution Control Standards; Notice of Decision Denying a Waiver of Clean Air Act Preemption for California's 2009 and Subsequent Model Year Greenhouse Gas Emission Standards for New Motor Vehicles, 73 Fed Reg. 12156, 12157 (March 6, 2008).

[14] GAO, *Clean Air Act: Historical Information on EPA's Process for Reviewing California Waiver Requests and Making Waiver Determinations*, GAO-09-249R (Washington, D.C.: Jan. 16, 2009).

[15] California State Motor Vehicle Pollution Control Standards; Notice of Decision Granting a Waiver of Clean Air Act Preemption for California's 2009 and Subsequent Model Year Greenhouse Gas Emission Standards for New Motor Vehicles, 74 Fed Reg. 32744, 32767, 32783 (July 8, 2009).

[16] Petition for Review, *Chamber of Commerce of the U.S. v. EPA*, No. 09-01237, D.C. Cir. (filed Sept. 8, 2009).

[17] For the remainder of this chapter, we refer to this joint rulemaking and related proposed standards as the proposed model year 2012 to 2016 rule.

[18] As required by EPCA, NHTSA must issue CAFE standards at least 18 months before they are implemented.

[19] The benefits and costs are estimated based on assumptions NHTSA and EPA made int he analyses, and may change if these assumptions (e.g., fuel price) change. See Appendix II for a discussion of NHTSA's analysis.

[20] Vehicle tailpipe emissions of carbon dioxide account for 90 to 95 percent of all vehicle GHG emissions.

[21] For model year 2008 through 2010 light trucks standards, manufacturers could opt to comply with the reformed footprint-based standards or an equivalent single fleetwide standard. Only General Motors opted to voluntarily comply with the reformed standard in 2008 and 2009. Starting with model year 2011 light trucks, all manufacturers must adhere to the footprint-based standard.

[22] Manufacturer compliance will be determined based on the fuel economy levels of actual vehicles produced compared with the CAFE footprint standard for each of those vehicles.

[23] The administration's goal has often been stated as a fleetwide average of 35.5 mpg. This value is equivalent to the 250 grams per mile carbon dioxide value if all of the carbon dioxide reductions come from fuel economy improvements.

[24] Some public comments on the Notice of Proposed Rulemaking suggested that NHTSA should mitigate against this possibility by imposing a "backstop"—a minimum CAFE standard that all manufacturers would be required to meet regardless of the footprint of their vehicles. EISA requires a backstop standard for domestically-manufactured passenger cars of either 27.5 mpg or 92 percent of the average projected fuel economy level of passenger cars in any given model year, whichever is greater. However, NHTSA did not include a backstop for imported passenger cars or light trucks in the September 2009 proposed rule.

[25] OMB Circular A-4, September 17, 2003.

[26] However, in the current proposed rule, NHTSA sought comment on providing manufacturers with CAFE credits for improving air conditioner efficiency for light trucks.

[27] This allowance is available during model years 2012 though 2015 to manufacturers whose vehicles sales in the U.S. in model year 2009 are below 400,000 vehicles.

[28] 42 USCS § 7522(a)(1).

[29] See GAO, *Vehicle Fuel Economy: Reforming Fuel Economy Standards Could Help Reduce Oil Consumption by Cars and Light Trucks, and Other Options Could Complement These*

Standards, GAO-07-921(Washington, D.C.: Aug. 2, 2007) and GAO, *Civil Penalties: Agencies Unable to Fully Adjust Penalties for Inflation under Current Law*, GAO-03-409 (Washington, D.C.: Mar. 14, 2003).

[30] 74 Fed. Reg. 49454, 49477 (Sept. 28, 2009).

[31] 42 U.S.C. § 7524(c)(2).

[32] Pub. L. No. 110-140, § 102(a).

[33] Rebound effect is the increase in vehicle miles traveled that result from the decreased costs of driving due to fuel economy increases.

[34] For example, while a 2002 National Academy of Sciences study estimated the costs of applying reduced rolling-resistance tires at $14 to $56, the work with the consulting firm found the cost to be a range of $6 to $9.

[35] However, the agency indicated in the proposed rule its intentions to refine its analysis for the final rule and sought comments to aid it in doing so.

[36] As single fleetwide CAFE standards (as opposed to attribute-based standards) can lead to lighter and smaller vehicles being sold due to their generally higher levels of fuel economy, these vehicles are also generally less safe than larger and heavier vehicles. NHTSA's analysis of safety effects of the proposed standards relies on the findings of a 2003 study that has been met with criticism by a number of experts and stakeholders because other studies have produced conflicting results regarding the relationship between vehicle weight, size, and safety. Some experts we met with cited the need for additional research on these issues given the lack of consensus, conflicting research, and the availability of new technology such as lightweight but durable materials for vehicle frames.

[37] This work is meant as an update to chapter 3 of the NAS study, *Effectiveness and Impact of Corporate Average Fuel Economy (CAFE) Standards* (National Academy Press, 2002).

[38] GAO, *Passenger Vehicle Fuel Economy: Preliminary Observations on Corporate Average Fuel Economy Standards*, GAO-07-551T (Washington, DC: Mar. 6, 2007).

[39] According to EPA staff, this study is still ongoing and the agency has expended $1.1 million to date. They plan to continue this study to evaluate additional GHG emissions- reducing technologies. EPA, *Light-Duty Technology Cost Analysis Pilot Study*, EPA-420-R09-020 (December 2009).

[40] EPA, *Automobile Industry Retail Price Equivalent and Indirect Cost Multipliers*, EPA- 420-R-09-003 (February 2009).

[41] EPA, *A Study of Potential Effectiveness of Carbon Dioxide Reducing Vehicle Technologies–Revised Final Report*, EPA-420-R-08-004a (June 2008).

[42] In addition, legislation may be enacted that would regulate GHG emissions from a wide range of sources on a national level, which could have an impact on CAFE and GHG emissions standards for vehicles. In June 2009, the House of Representatives passed such a bill—H.R. 2454, 111th Cong. (2009). The Senate is currently considering similar legislation in S. 1733, 111th Cong. (2009).

[43] GAO, *Health and Safety Information: EPA and OSHA Could Improve their Processes for Preparing Communication Products*, GAO-08-265 (Washington, D.C.: Mar. 31, 2008).

[44] In 2007, GAO reported on concerns with NHTSA's analysis in setting CAFE standards. Specifically, we found that experts were concerned about the values used for certain inputs, such as the estimated social cost of carbon dioxide emissions, that NHTSA officials used in the computer model maintained by DOT's Volpe Center. See GAO 07-921.

[45] These groups are engine, transmission, electrification/accessory, hybrid, or vehicle.

[46] Ground-level ozone, particulate matter, carbon monoxide, nitrogen oxides, sulfur oxides, and lead are called criteria pollutants because EPA regulates them by developing human health-based or environmentally-based criteria (science-based guidelines) for setting permissible levels.

[47] Specifically, NHTSA used product plans to obtain estimates of the volume of each vehicle model a manufacturer expects to produce for sale in future model years, as well as detailed

information on the characteristics of individual vehicle models including engines, transmissions, and other technology.

[48] See 49 CFR Part 512.

[49] The same vehicle baseline was used in EPA's OMEGA model.

[50] For example, as reported in the proposed rulemaking, one manufacturer's product plans contained important errors in estimates of vehicle footprints.

[51] A conventional hybrid uses both gasoline and energy stored in a battery to power the vehicle. A plug-in hybrid can be plugged into an electrical outlet to charge a battery that can power the vehicle.

[52] In addition, NHTSA solicited public comments in the Notice of Proposed Rulemaking on the methodology and data sources used to develop the baseline vehicle fleet and the reasonableness of the results.

[53] These sources include the 2002 NAS report, the 2004 Northeast States Center for a Clean Air Future study, the 2004 California Air Resources Board Initial Statement of Reasons in support of their carbon rulemaking, a 2006 study by Energy and Environmental Analysis for DOE, a study by the Martec Group for the Alliance of Automobile Manufacturers and the 2008 update to that study, and vehicle fuel economy certification data. Both agencies also reviewed published technical literature that addressed the issue of CO_2 emission control and fuel economy, such as papers published by the Society of Automotive Engineers and the American Society of Mechanical Engineers. In addition, confidential data submitted by vehicle manufacturers in response to NHTSA's request for product plans, and confidential information shared by automotive industry component suppliers in meetings with EPA and NHTSA staff held during the second half of the 2007 calendar year were used as a cross-check of the public data mentioned above but not as a significant basis for the proposed model year 2012 through 2016 rule.

[54] Turbocharging and downsizing reduces an engine's pumping losses at lighter loads in comparison to a larger engine by increasing the rate at which the engine is able to draw air into the engine's combustion chambers.

[55] A teardown study is a study in which a vehicle is disassembled in order to determine the specifications of its components, including their costs.

[56] Cylinder deactivation can improve the efficiency of the engine by disabling or deactivating (usually) half of the cylinders when the load is less than half of the engine's total torque capability. In cylinder deactivation, the valves are kept closed, and no fuel is injected. As a result, the trapped air within the deactivated cylinders is simply compressed and expanded as an air spring, with reduced friction and heat losses. The active cylinders combust at almost double the load required if all of the cylinders were operating. Pumping losses are significantly reduced as long as the engine is operated in this "part-cylinder" mode.

[57] *Center for Biological Diversity v. National Highway Traffic Safety Administration*, 508 F.3d 508 (9th Cir. 2007).

[58] EIA's main price projection is its Reference Case, which represents EIA's current judgment regarding exploration and development costs and accessibility of oil resources in countries that are not members of the Organization of the Petroleum Exporting Countries (OPEC).

[59] In the EIA's 2009 Annual Energy Outlook, the High Oil Price Case uses a range of prices from $3.36 in 2011 to $5.47 in 2030, and the Low Oil Price Case uses a range of prices from $2.19 in 2011 to $2.04 in 2030.

[60] For example, in its analysis for the model year 2008 through 2011 light truck rule, NHTSA proposed using EIA's reference price case but received comments critical of this decision in light of retail fuel prices that were significantly higher than EIA's reference case. For the final model year 2008 through 2011 light truck rule, NHTSA considered the comments the agency had received and decided to use the EIA's high price case to more accurately estimate the trajectory of gas prices in the future.

[61] According to the proposed rule, standards set based on maximizing net benefits would reach an estimated 40.9 mpg fleet average in model year 2016.

[62] Rebound effect is the increase in vehicles miles traveled that result from the decreased costs of driving resulting from fuel economy increases.

[63] While monetized values for safety with respect to loss of life are not included as an input in the Volpe model, safety impacts are considered in determining the appropriate level for CAFE standards.

[64] Charles J. Kahane, *Vehicle Weight, Fatality Risk and Crash Compatibility of Model Year 1991-99 Passenger Cars and Light Trucks*, DOT HS 809 662, October 2003.

[65] R.M. Van Auken and J.W. Zellner, Dynamic Research, Inc., *Supplemental Results on the Independent Effects of Curb Weight, Wheelbase, and Track on Fatality Risk in 1985-1998 Model Year Passenger Cars and 1985-1997 Model Year LTVS*, DRI-TR-05-01 (Torrance, Calif., May 2005).

[66] GAO, *Reexamining Regulations: Opportunities Exist to Improve Effectiveness and Transparency of Retrospective Reviews*, GAO-07-791 (Washington, D.C.: July 16, 2007).

[67] See for example "Comparison of EPA and Other Estimates of Mobile Source Rule Costs to Actual Price Changes", J. Anderson and T. Sherwood, US EPA, published in the Society of Automotive Engineers Technical Paper Series, SAE 2002-01-1980, 2002.

[68] As we discussed earlier, there is continuing controversy over the relationship between vehicle size, weight, and safety. Some experts we met with said that some manufacturers, in order to meet increased CAFE standards, may keep vehicle footprint constant while reducing overhang (the area of the car ahead of the wheelbase), which could make passengers more vulnerable in crashes. A retrospective analysis could help determine the extent to which this occurs and the potential safety implications of a footprint-based CAFE standard even assuming that footprints are not reduced.

[69] NHTSA does, however, produce an annual Summary of Fuel Economy Performance report that provides information on CAFE standards in the previous year and the actual fuel-economy level of all manufacturers that are subject to the standards. In addition, they publish a summary of CAFE fines assessed on an annual basis.

[70] In the proposed rule, NHTSA used 39 technology applications, such as engine turbocharging/downsizing, variable valve timing, cylinder deactivation, and engine friction reduction.

[71] In previous rules, NHTSA has relied on confidential product plans provided by manufacturers to create the baseline fleet, but it has shifted away from that approach to make the baseline data more transparent for the proposed rule.

[72] For the purpose of applying technologies, NHTSA distinguishes vehicles by subclass, including subcompact car, subcompact performance car, compact car, compact performance car, midsize car, midsize performance car, large car, large performance car, minivan, small SUV/pickup/van car, midsize SUV/pickup/van, and large SUV/pickup/van.

[73] Learning is a means of capturing the reduction in cost of the components and manufacturing process involved with a technology. A reduction in cost takes place when the volume of deployment of that technology increases dramatically (volume-based) or when reasons related to other factors, such as negotiated contractual agreements between suppliers and original equipment manufacturers, occur over a period of time (time-based).

[74] Fleet penetration is the percentage of a fleet that a technology can be applied to in a given model year, which is based on supply constraints and other reasons.

[75] The five technology groups are engine, transmission, electrical accessory, hybrid, or vehicle.

[76] The "next-best" methodology operates as follows: the Volpe model considers technologies within one of the five technology groups in sequential order, proceeding to the next technology if the phase-in cap has been reached for a particular technology (i.e., 85 percent penetration for turbocharged/downsized engines). The Volpe model determines whether the technology can be applied to any set of vehicles, evaluates the effective cost of doing so, and identifies the technology from each technology group that would yield the lowest effective cost.

[77] Ozone, particulate matter, carbon monoxide, nitrogen oxides, sulfur dioxide, and lead are called criteria pollutants because EPA regulates them by developing human health-based or environmentally-based criteria (science-based guidelines) for setting permissible levels.

CHAPTER SOURCES

The following chapters have been previously published:

Chapter 1 – This is an edited, excerpted and augmented edition of a United States Congressional Research Service publication, Report Order Code R40506, dated December 9, 2009.

Chapter 2 – This is an edited, excerpted and augmented edition of a United States Congressional Research Service publication, Report Order Code R40984, dated December 15, 2009.

Chapter 3 – This is an edited, excerpted and augmented edition of a United States Congressional Research Service publication, Report Order Code RS22870, dated February 18, 2010.

Chapter 4 – This is an edited, excerpted and augmented edition of a United States Congressional Research Service publication, Report Order Code R40168, dated February 4, 2010.

Chapter 5 – This is an edited, excerpted and augmented edition of a United States Government Accountability Office (GAO), Report to the Chairman, Subcommittee on Energy and Environment, Committee on Energy and Commerce, House of Representatives. Publication GAO-10-336, dated February 2010.

INDEX

A

B

C

D

E

F

G

H

I

T

U

V

W